Advances in Bioelectrochemistry Volume 2

Frank Nelson Crespilho
Editor

Advances in Bioelectrochemistry Volume 2

Biomimetic, Bioelectrocatalysis and Materials Interfaces

Springer

Editor
Frank Nelson Crespilho
University of Sao Paulo
São Carlos, São Paulo, Brazil

ISBN 978-3-030-95269-3 ISBN 978-3-030-95270-9 (eBook)
https://doi.org/10.1007/978-3-030-95270-9

This Springer imprint is published by the registered company Springer Nature Switzerland AG
The registered company address is: Gewerbestrasse 11, 6330 Cham, Switzerland

Contents

Biomimetics Applied in Electrochemistry

Iago A. Modenez

Abstract Nature has evolved through billions of years to achieve highly efficient biological processes and materials with properties that go, most of the time, far beyond the current human know-how. Not only to deeply understand these selective systems, but also to gain inspiration for the design of new processes and materials with similar behavior, the principles of biomimicry and bioinspiration have been applied by the scientific community and the industry. The general strategy for that is firstly, through scientific analysis, systematize the fundamental mechanisms that underlie a particular biological process, and then apply these concepts in the fabrication of novel biomimetic/bioinspired materials with enhanced performance. The inspiration coming from natural sources can be structure-wise and/or function-wise and, in the recent years, has had a significant impact on the bioelectrochemistry field, as most of the available electrochemical techniques are useful tools to investigate these systems. This chapter discusses the recent trends in biomimicry and bioinspiration in the bioelectrochemistry field, mainly focusing on biomimetic membranes, reconstituted membrane proteins, protein-based electrodes, biomimetic enzymes, genetic materials, and live cells.

1 Introduction

Nature expresses the highest level of complexity with structures and functions that are remarkable in terms of selectivity and efficiency. The biological entities and processes have evolved through billions of years to perform a variety of functions including catalysis, sensing, light-harvesting, molecular recognition, self-assembly, charge transfer, structural support, signal transduction, and the list keeps going on [1]. Taking advantage of this evolutionary experimentation to learn and gain inspiration can be a helpful, but challenging, approach to address many problems faced by humanity. These are fundamental aims of the bioinspiration field, and ultimately, of

I. A. Modenez (✉)
iSm2 UMR 7313, CNRS, Aix-Marseille Université, 13397 Marseille, France
e-mail: iago_modenez@alumni.usp.br

F. N. Crespilho (ed.), *Advances in Bioelectrochemistry Volume 2*,
https://doi.org/10.1007/978-3-030-95270-9_1

the biomimetics (a term that is derived from the Greek word "bios" meaning "life" and "mimesis" meaning "to imitate") [2, 3].

Although the formal distinction between biomimicry and bioinspiration is not clear-cut, the current definition of biomimicry involves the direct replication/imitation of the structure–function relations observed in biological systems, while bioinspiration is mostly concerned in firstly revealing structure–property mechanisms and formulating systematic theories and then, applying the principles that underlie natural processes to non-biological systems, pushing these principles to higher levels and going beyond what nature offers [4, 5]. Furthermore, in a simplified way, we can say that a biomimetic material—structure-wise and/or property-wise—is the result of observation and study of the structure and function taking biological systems as a model [6]. Interestingly, over the last two decades, the number of publications related to biomimetics has increased from around 100 research articles in 1995 to more than 2500 research articles published only in 2017 [7]. This demonstrates not only the increasing interest of the scientific community in the field, but also an evolution in the understanding of biological systems at a molecular scale, in the wide range of applications—from chemistry [8], medicine [9], material science [3, 10], energy [11] to robotics [12], architecture [13], and arts [14]—and in the techniques and methods used to investigate these systems [15].

Electrochemical techniques have been an important tool to study biomimetic and bioinspired materials in the field of bioelectrochemistry. Monitoring electron transfer (ET) processes, current flow, and energetic state modulation due to the application of a certain potential is very insightful and makes it possible to gather valuable information regarding these systems [16]. Moreover, the combination of electrochemical techniques and surface-sensitive methods, such as spectroscopic, spectrometric, and microscopic techniques, has been employed to better understand the dynamics and structure of biomimetic systems [17].

In this chapter, it is described the recent trends in the field of biomimicry and bioinspiration, focusing mostly on the bioelectrochemistry field. All the biomimetic or bioinspired materials, in terms of structure and/or properties, presented here have an equivalent biological inspiration, which is also introduced. In addition, the electrochemical and coupled techniques (in situ or operando techniques) applied to study these systems are briefly discussed. The first section discusses the importance of biomimetic membranes and reconstituted membrane proteins in the understanding of the relationship between the cells and the surrounding environment. The following section is focused on protein-based electrodes and biomimetic enzymes for biocatalysis, followed by a section in which genetic materials and living organisms are used to construct biomimetic materials. Finally, broader applications—especially concerning energy storage materials and bioelectrocatalysis—are presented, along with the final conclusions and some future perspectives in the field.

2 Biomimetic Membranes

Biological membranes, which form the outer boundary of living cells allowing a selective communication between their interior and the surrounding environment [18], are involved in several crucial processes, including signaling, cell–cell recognition, active transport of molecules, transmembrane ion-conducting channels, compartmentalization and adhesion processes, and others [19]. They can also incorporate a large variety of proteins, carbohydrates, cholesterols, redox molecules, and oligosaccharides that play vital roles especially in redox reactions, such as in photosynthesis and respiration. The biological membranes' high complexity in composition and structural scaffold has led to the development of several simplified membranes that can be used as biomimetic models—as some of their physical–chemical properties are most likely similar to natural cell membranes—to gather a deep knowledge into the relation between membrane structure and cellular function.

The formation of biomimetic membranes at metal electrode surfaces is quite interesting as it allows the combination of electrochemical techniques (i.e., cyclic voltammetry (CV), chronocoulometry, electrochemical impedance spectroscopy (EIS), and differential capacitance) and surface-sensitive methods (i.e., spectroscopy, microscopic methods, and neutron scattering) to characterize the systems, in terms of both electrochemical properties and molecular level information. Moreover, by depositing biomimetic membranes at gold surface electrodes, for instance, it is possible to directly associate changes taking place in the membrane with the electrical signal as current, resistance, and capacitance changes, which ultimately contribute to the development in the biosensing and biomedical devices field [20]. Among the different types of biomimetic membranes that can be formed at metal surfaces, it is possible to highlight the supported lipid monolayers (sLMs), supported planar bilayer lipid membranes (sBLMs), tethered bilayers (tBLMs), and floating bilayer lipid membranes (fBLMs) [21, 22], which are usually built by combination of Langmuir–Blodgett and Langmuir-Schaefer (LB-LS) transfer methods or by vesicle fusion (VF).

The supported lipid monolayers (sLMs) are the biomembranes' simplest models, firstly studied by Plank and Miller using mercury as the substrate [23]. This model has found several practical applications ranging from structure-dependent interactions with flavonoids molecules [24], screening of organic pollutants in water [25] and interactions with nanoparticles [26] to investigations concerning the adhesion of cells to charged electrode surfaces [27]. Furthermore, coupled electrochemical techniques, such as electrochemical scanning tunneling spectroscopy (EC-STM) and electrochemical atomic force microscopy (EC-ATM), can be used, for example, to better understand and visualize the formation mechanism of lipid vesicles on an Au(111) surface under electrochemical control [28], potential-dependent structural transitions, and hydration/dehydration processes triggered by an electric field in chitosan-derived molecular films [29]. In addition, Prieto and coworkers [30] reported the use of CV and EIS techniques to analyze the stability and electrical properties of a sLM on gold electrode, as well as, the release mechanism of an anticancer drug.

The first generation of lipid bilayers models supported on a solid substrate is the supported planar bilayer lipid membranes (sBLMs). In this system, the inner layer is directly adsorbed onto the substrate (such as Au, Ag, or mica) while the outer leaflet is exposed to the electrolyte; therefore, the biomembrane is exposed to an asymmetric environment. A number of studies have been carried out in order to better understand the formation mechanism [31], the potential-dependent structural transitions and permeability changes of sBLMs supported on gold surfaces [32], as well as their interactions with antibiotic peptides [33, 34].

This type of biomimetic membrane finds applications in a wide range of fields, including biosensing and bioelectronic devices, as reported by Su and coworkers [35]. The authors present a method for generating sBLMs on a conducting polymer (poly(3,4-ethylenedioxythiophene) polystyrene sulfonate; here named as PEDOT:PSS) device using a solvent-assisted lipid bilayer (SALB) technique inside a microfluidic channel. This technique consists in suspending lipid micelles in an isopropanol/water mixture that is exposed to the surface of interest, in this case the PEDOT:PSS layer. During the incubation time, the micelles adsorb to the surface, then, with a slow replacement of the solvent with increasing aqueous buffer fraction; the lipid micelles rapidly self-assemble into a planar bilayer lipid membrane at the substrate surface (Fig. 1A). Two different model membranes were constructed following the method: a mammalian and a bacterial biomimetic membrane. Bilayer formation on a PEDOT:PSS/Au electrode surface was further validated by EIS measurements (Fig. 1B), and its sensing capability was probed by the interaction between the mammalian model with an α-hemolysin (α-HL), a pore-forming toxin from *Staphylococcus aureus*, and antibiotic interactions between the bacterial model with polymyxin B (PMB), a cationic peptide used to treat Gram-negative bacterial infections.

When the supported lipid bilayer assembles, there is a shift in the phase plots (Fig. 1C, E, black and red dots) for both biomimetic membranes due to the formation of an additional layer on the electrode, which also generates the characteristic semicircle in the Nyquist plot, indicating an RC element (Fig. 1D, F, black and red dots). The mammalian and bacterial membrane models showed a resistance of 0.54 ± 0.05 and 0.46 ± 0.09 kΩ cm^2, respectively, and a calculated capacitance of 376.6 ± 166.1 and 950.0 ± 808.7 μF cm^{-2}, respectively. The authors attributed the higher capacitance value for the bacterial membrane model to an ionic reservoir due to additional electrostatic interactions between the negatively charged lipids domains and the negatively charged PSS regions on the polymer surface. Furthermore, the integrity of the membrane models was evaluated by the changes in the resistance upon interaction with a toxin and an antibiotic compound. It was reported a $23.2\% \pm 17.8\%$ decrease in the resistance of the mammalian membrane upon interaction with α-HL (Fig. 1D, blue dots), while for the bacterial membrane model, the calculated decrease was $99.63\% \pm 1.86\%$ with both the phase and Nyquist plot resembling the lipid-free state (Fig. 1F, blue dots), indicative of a possible destruction of the lipid bilayer upon interaction with PMB.

In addition, a set of electrochemical techniques, such as differential capacitance, CV, chronocoulometry, differential pulse voltammetry (DPV), and EIS, has been

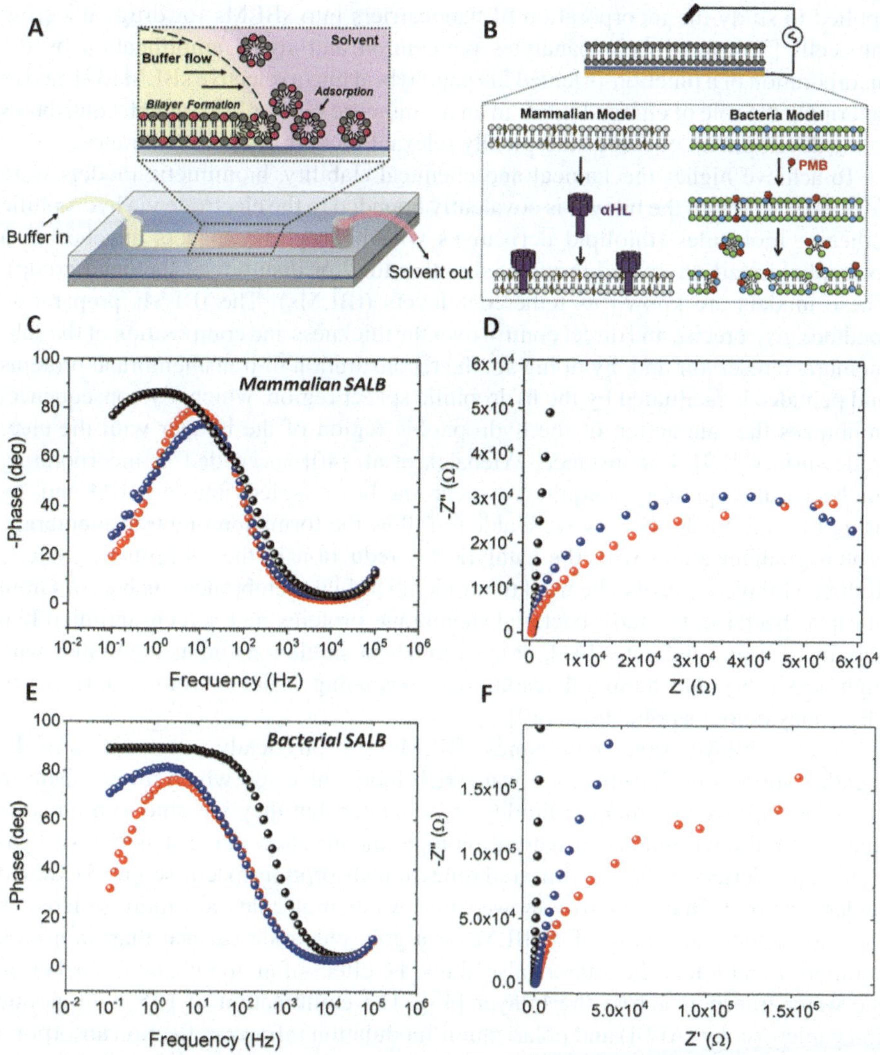

Fig. 1 SALB formation inside a microfluidic channel. A The magenta color denotes the solvent phase containing lipid micelles while the yellow represents aqueous buffer flowing-in to replace the solvent and forming the supported bilayer at the channel surface; **EIS configuration used in the paper. B** The thin blue layer represents the PEDOT:PSS film and the dark yellow layer represents the gold contact supporting the polymer and planar bilayer. Below the diagram are cartoon representations of the two bilayers formed with the method, and the toxin (α-HL) or compound (PMB) used to disrupt them. POPC (light gray) and cholesterol (yellow) are the components of the mammalian bilayer while POPE (green) and POPG (blue) are the components used for the bacterial bilayer. **Impedance monitoring of the SALBs on PEDOT:PSS/Au electrodes.** Phase of the impedance and the corresponding Nyquist plot for the mammalian membrane model (**C, D**) and bacterial membrane model (**E, F**). Before the SALB addition (black), after the SALB addition (red), after the insertion of the toxin α-HL (5×10^{-6} M) or PMB (0.5 mg mL^{-1}) (blue) in the mammalian and in the bacterial bilayers, respectively. Reprinted with permission from [35], copyright 2019 American Chemical Society

applied to study the incorporation of nanocarriers into sBLMs for drug delivering into cells [36], to probe bioanalytes recognition and signal amplification by the incorporation of a function-oriented 2D nanohybrid into a selective sBLM [37], and to determine the role of charged lipids in an asymmetric bilayer [38], which contributes to the development of more biologically relevant models of cell membranes.

To achieve higher mechanical and chemical stability, biomimetic models were developed in which the bilayer is covalently bounded to the electrode via hydrophilic tethering molecules (thiolipid derivatives with hydrophilic spacers attached to a hydrophobic tail group and terminated with a thiol or disulfide at the head group). These models are known as tethered bilayers (tBLMs). The tBLMs preparation methods give precise and direct control over the thickness and composition of the sub-membrane reservoir and, by doing so, the reconstitution of transmembrane proteins and peptides is facilitated by the hydrophilic spacer region, which, by consequence, minimizes the interaction of the hydrophobic region of the bilayer with the electrode surface [39]. For instance, Wiebalck et al. [40] succeeded in incorporating the bacterial respiratory ubiquinol/cytochrome bo_3 coupleda into a tBLM and, by using EIS and SEIRAS, they were able to follow the formation of a transmembrane proton gradient along with the catalytic O_2 reduction in the biomimetic system. Tethered bilayers can also be used to mimic microbial membranes, such as of Gram negative bacteria, to study bacterial membrane proteins and screen antimicrobial activity of drug candidates [41], or even to detect single ion channel activities with high sensitivity and temporal resolution envisioning nanoelectronic interfaces to electrophysiology applications [42].

Floating bilayer lipid membranes (fBLMs) are physically separated from the metallic support ($\sim$2.4 nm) by a water-rich lubricant layer, which allows them to show an improved membrane fluidity in a manner that they are able to mimic the quasi-natural environment of natural biologic membranes. For instance, investigations using surface-enhanced infrared reflection absorption spectroscopy (SEIRAS) under electrochemical control showed that water molecules are more ordered in the sub-membrane region of a fBLM on a gold electrode surface than in a bulk solution. In addition, the authors also show the effect of an ion channel blocker on the water transport across the bilayer [43]. The combination of EIS with atomic force microscopy (AFM) and polarization-modulation infrared reflection absorption spectroscopy (PM-IRRAS) measurements is a powerful approach to visualize the formation of conducting pores in a fBLM due to the insertion of an antimicrobial peptide, as reported by Abbasi et al. [44], as well as the inhibition of ion transport thought membrane channels due to the action of amiloride drug [45].

2.1 Reconstituted Proteins into Biomimetic Membranes

The development and optimization of biomembrane models are quite important not only to understand the formation and dynamics of biological membranes themselves, but specially concerning reconstituted protein-based biomimetic platforms, which find applications in a wide range of fields, coming from molecular interactions studies, materials development and biosensors to energy harvesting, medicine, and water purification [46]. Membrane proteins are biomolecules with high specificity and selectivity yield, but when they are not in their native-like environment, they often go through denaturation processes leading to their inactivation, which makes the electrochemical characterization and direct applications of these proteins limiting at a first sight. One good strategy then is to incorporate membrane proteins in model lipid bilayers that mimicry their natural environment and attach this system to the electrode surface. For instance, small ion channel-forming proteins, such as gramicidin A and α-hemolysin, have been successfully incorporated into lipid bilayers for ionic and molecular recognition and quantification [47, 48], while the pH-depended transporting properties of a complex multispan membrane protein, designated Bot1, inserted into a tBLM were probed using EIS [49]. A tBLM model was also applied to study the kinetics of secondary structure conformational changes occurring in cytochrome c oxidase (CcO) under electrochemical control, using time-resolved surface-enhanced IR-absorption spectroscopy (tr-SEIRAS) [50]. In this case, the use of a biomimetic system has two major advantages: it allows the direct ET between the CcO and the electrode without the need of mediators and, moreover, due to the geometry of the tBLM, a hydrophilic layer between the electrode and the lipid membrane allows the electrochemically reduction of protons (resulting from the enzyme activity) to molecular hydrogen.

The reconstitution of proteins into biomimetic membranes is not always achievable, since the protein, in some cases, can induce significant damage in the membrane, hampering the incorporation, as reported by Karaballi et al. [51]. In this study, the authors monitored the interaction between a model amyloid-forming protein and a sBLM, formed on a nanostructured substrate by LB-LS deposition. Electrochemical surface-enhanced Raman spectroscopy (EC-SERS) was used to monitor molecular level changes resulting from the interaction between the protein, at different stages of aggregation, and the sBLM, in which the latter was significantly deteriorated by the protein oligomers and protofibrils.

The ability of harness and transduce energy, which is linked to the vectorial transfer of ions across the cell membrane, is a notorious strategy in nature. The energy stored as electrochemical gradient is, then, used to carry out several biochemical processes, including the chemical synthesis of ATP driven by the enzyme F_1F_0-ATPase. The development of biomimetic devices capable of reproducing nature's behavior is a significant challenge. In this sense, Gutiérrez-Sanz and coworkers firstly demonstrated that a proton gradient can be produced by electroenzymatic H_2 oxidation on an electrode surface covered by a phospholipid bilayer (PBL) [52]. The authors reported that this gradient can be harnessed in an artificial system to fuel ATP synthesis on an Au electrode modified to include a membrane-bound NiFeSe hydrogenase (Hase) and a F_1F_0-ATPase in the presence of a PBL [53] (Fig. 2A). Initially, the biomimetic construction was followed using ATM enabling the visualization of the NiFeSe hydrogenase monolayer due to its covalent binding to the surface modified with a self-assembled monolayer (SAM) of 4-aminothiophenol, as well as the protrusions occurring on top of the bilayer corresponding to the membrane-embedded ATPase.

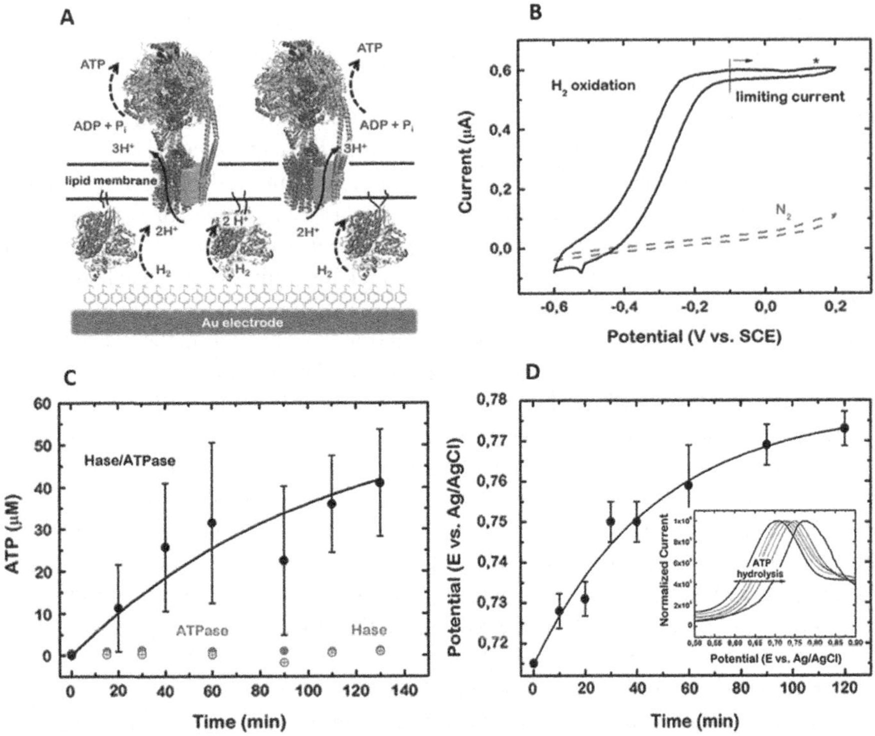

Fig. 2 Representation of the supramolecular construction to synthesize ATP from enzymatic H_2 electrooxidation. A Hase immobilized covalently on a SAM-modified (4-aminothiophenol) Au electrode and anchored to a phospholipid bilayer (PBL), in which it is embedded F_1F_0-ATPases enzymes. The biomimetic system uses the protons produced from H_2 electrooxidation to synthesize ATP; **Cyclic voltammetry and ATP production. B** Cyclic voltammograms of the Hase/PBL/ATPase-modified electrode before activation under N_2 (dashed line) and after Hase activation through H_2 incubation (solid line) (0.1 mol L^{-1} phosphate buffer pH 8.0, scan rate = 0.01 V s^{-1} and T = 30 °C); **C** ATP synthesis from ADP + P_i at 150 mV *vs.* SCE under 1 atm H_2 (0.1 mol L^{-1} phosphate buffer pH 8.0). ATP concentration is shown as a function of time for Hase/PBL/ATPase (black solid circles), PBL/ATPase (gray solid circles), and Hase/PBL (gray open circles) electrodes. The error bars were estimated from the standard deviation of 3 measurements; **Proton gradient. D** Simultaneous proton pumping activity of F_1F_0-ATP synthase over time monitored by DPV (inset). The oxidation peak of the SAM on Au is shown as a function of time after ATP hydrolysis started. The error bars were estimated from the standard deviation of 2 measurements. Reprinted by permission from [53], copyright Wiley–VCH 2016

To check the bioelectrocatalytic activity of the modified electrode with both enzymes, the Hase was activated in an anaerobic chamber under 1 atm of H_2 and subsequently a cyclic voltammogram was obtained sweeping from −0.6 to + 0.2 V, in which it is observed an increasing anodic current until it reaches a plateau at −0.2 V attributed to the oxidation of H_2 to protons (Fig. 2B). In a second experiment, a continuous potential of +150 mV (*vs.* SCE) was applied during 130 min in the presence of 500 μM of ADP and under 1 atm of H_2, in order to generate protons at the

PBL/electrode interface of the biomimetic device. The experiments showed that ATP is produced during hydrogen oxidation catalyzed by the immobilized Hase only in the presence of F_1F_0-ATPase, ADP, and phosphate. Moreover, ATP was no detected in control experiments performed with just one of the enzymes immobilized on the electrode surface (Fig. 2C). The capacity of the F_1F_0-ATPase incorporated into the PBL to hydrolyze ATP and to generate a proton gradient in the absence of Hase was then investigated by the authors. As the SAM present on the electrode surface has its redox potential dependent on the media pH, it can act as a pH-sensitive probe. The results showed that, after incubation of the electrode in an ATP solution, a shift in the oxidation peak of the 4-aminothiophenol measured by DPV was attributed to the pH change at the PBL/electrode interface associated with the ATP hydrolysis by the reconstituted F_1F_0-ATPase (Fig. 2D). This electrode-assisted conversion of H_2 into ATP could be applied in the future to locally generate this biochemical fuel in biomedical devices or even in valuable products' enzymatic synthesis.

Not only the synthesis, but also the detection of ATP is important, as this molecule is a good indicator of cell metabolism, it can reveal a possible presence of microbial colonies, which is an important issue in food industry and in healthcare settings, for instance. García-Molina et al. [54] reported a potentiometric biosensor based on an ATPase reconstituted in a fBLM over Au electrodes modified with a 4-aminothiophenol SAM. The biosensor provided a quantified value of ATP concentration in the bulk within 5–10 min ranging from 1 μM to 1 mM.

Natural photosynthetic processes, such as photosynthesis, have been an inspiration for the development of biomimetic materials in the energy harvesting field. Photosystem I (PSI), a nano-scale biological photodiode, is a transmembrane protein complex responsible for the photoactivated charge separation with near unity quantum efficiency during photosynthesis. To fabricate stable bio-hybrid devices capable of high photocurrent generation, it is important to systematically assemble well-oriented and functional PSI onto the desired bio-abio interfaces through suitable biomimetic protein scaffoldings. Niroomand et al. [55] investigated the photoactivity of biomimetic constructs of cyanobacterial PSI reconstituted within a negatively charged sBLM—mimicking the natural thylakoid membrane—assembled on an SAM Au electrode. AFM imaging and force spectroscopy were used to probe the incorporation of PSI into the sBLM, while electrochemical measurements (in particular, chronoamperometry) indicated up to fourfold to fivefold enhancements in photocurrent densities generated from the biomimetic system when compared to the measurements from a dense monolayer of individual PSI on SAM/Au substrates. The authors stated that these findings contribute to the development of photochemical energy conversion and artificial photosynthetic systems, as the fundamental understanding of the optoelectronic properties of PSI under biomimetic reconstitution is key to achieve optimized devices.

The increase in biocatalytic activity of proteins and enzymes can be obtained by their incorporation into biomimetic lipid membranes, as already seen in the previous examples, but also another valuable approach is using multilayered lipid membrane stacks that show various functional advantages compared to single-lipid membranes, such as increased surface area, compartmentalization of biomolecules, and more spatially organized processes, as reported by Heath and coworkers [56]. The authors engineered a lipid multilayer matrix assembling two different membrane proteins, in which poly-L-lysine electrostatically links the negatively charged sBLM on an Au electrode. In a first set of experiments, the ubiquinol oxidase, cytochrome bo_3 (cyt bo_3), was reconstituted in the lipid assemble (Fig. 3A). The catalytic activity

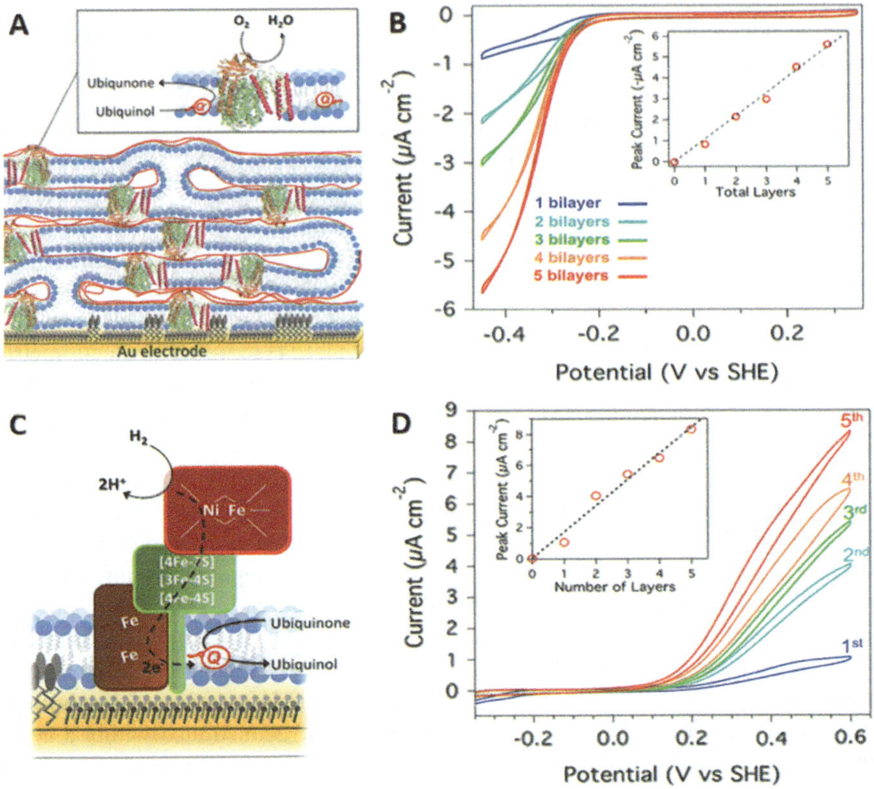

Fig. 3 Representation of the possible interconnected multilayer membrane structure containing cyt bo_3. A The ubiquinone molecules are reduced to ubiquinol at the Au electrode surface, which is then oxidized by cyt bo_3, catalyzing the reduction of O_2 to H_2O; **B** cyclic voltammograms of increasing bilayer stacks containing cyt bo_3. Inset depicts the catalytic current measured at −450 mV *vs.* SHE as function of the bilayers number (air saturated buffer (~250 × 10^{-6} M O_2) was used as supporting electrolyte + 20 × 10^{-5} M of 3-(N-Morpholino) propanesulfonic acid + 30 10^{-3} M of Na_2SO_4, pH 7.4, T = 20 °C at scan rate of 10 mV s^{-1}). **Illustration of MBH activity on the membrane-modified electrode. C** The electrons generated from H_2 oxidation are used to reduce ubiquinone by the cyt b_{562} subunit (shown in brown) to ubiquinol. Ubiquinone is then reoxidized at the electrode surface; **D** cyclic voltammograms of increasing bilayer stacks containing MBH. Inset depicts the catalytic current measured at +600 mV *vs.* SHE as function of the bilayers number (supporting electrolyte = 5% H_2, 95% N_2 saturated buffer + 20 × 10^{-3} M of 3-(N-Morpholino) propanesulfonic acid + 30 10^{-3} M of Na_2SO_4, pH 7.4, T = 30 °C at scan rate of 10 mV s^{-1}). Reprinted by permission from [56], copyright Wiley–VCH 2017

measured by CV showed that, at low electrode potentials (<0.2 V), quinone is reduced to quinol at the electrode surface and diffuses to cyt bo_3 where it is oxidized by the enzyme, coupled to the reduction of O_2 to H_2O.

Moreover, increasing the number of layers, which by consequence also increases the concentration of enzyme, leads to a ~2.5-fold increase in the catalytic current (Fig. 3B). The same behavior was observed when a membrane-bound NiFe hydrogenase (MBH) was incorporated in the biomimetic lipid membrane (Fig. 3C). The electrons coming from the oxidation of H_2 to protons are transferred from the MBH to the electrode through the quinone pool, as in the natural biological process. Each additional bilayer increases the catalytic current, as shown by the CVs (Fig. 3D). These results hold substantial technological potential for biomimicry, finding applications in the design of novel protein arrangements for photonics, catalysis, and sensing, for instance.

3 Bioinspired Protein-Based Electrodes and Biomimetic Enzymes

The combination of protein and redox enzymes—which have the ability to catalyze several reactions applicable for energy, biomedical, and environmental purposes—and abiotic materials, such as the electrodes, have attracted researchers' interests in the past years [57, 58]. The appropriate electron coupling and long-term stability of these systems remain a fundamental challenge until nowadays. More recently, these challenges have been addressed through biomimetic and bioinspired approaches envisioning the combination of electrodes with specific redox potentials and flexibility, along with highly stable and properly oriented redox proteins. The construction of these protein-based electrodes is based on borrowing structures and strategies found in biological ET systems in order to understand and design improved engineered electrode devices [59].

One critical step in the construction of bioinspired protein-based electrodes is the immobilization/stabilization of the protein at the electrode interface, which can be accomplished by electrostatic binding, energy-based immobilization, and covalent binding. The electrostatic immobilization methods involve tailoring protein binding to interfaces based on the opposite charges between enzyme and electrode [60]. This approach mimics the structure and chemical nature of protein binding sites, and the formation of a donor–acceptor complex enables protein–protein electron transfer. Another similar approach is binding of protein by interface chemical modification, in which the binding is not based on electrostatic interactions but on the protein's attraction to high-energy chemical groups (-OH, -COOH, for instance) [61]. Since most of the redox cofactors are covalently stabilized within the polypeptide chain in proteins, a covalently bounding approach can also be used in bioelectrochemical devices, specially using polymeric interfaces through cross-linking on electrodes.

The step of immobilization of the biomolecule at the electrode surface is quite important as many other parameters, such as protein loading and long-term stability, will be directly influenced by this. Protein orientation at the electrode surface, for instance, is crucial to achieve maximum electron transfer. This point becomes even more important when two different enzymes are immobilized on the interface.

Efrati and coworkers [62], inspired by the natural photosynthetic apparatus, reported the construction of a photo-bioelectrochemical cell for the conversion of solar light energy into electrical power. The main idea was to conjugate a redox enzyme to a photosystem-based photoelectrochemical electrode to generate photocurrents with the concomitant driving of an enzyme-catalyzed reaction. As shown in Fig. 4A, the photoexcitation of photosystem I (PSI) leads to the transferring of electrons to the electrode surface through a relay unit (R_1), whereas the PSI oxidizes the redox enzyme by the ET mediator (R_2), resulting in the biocatalyzed oxidation of the substrate. In order to produce a directional electron flow to an indium tin oxide (ITO) glass electrode, the authors set out a biomimetic system in which the communication between PSI and glucose oxidase (GOx) was achieved using a pyrroloquinoline quinone (PQQ) monolayer and an osmium redox polymer as relay units (R_1 and R_2, respectively), as shown in Fig. 4B. Considering the measured redox potentials of the PQQ monolayer, the osmium redox polymer, and the PSI, a vectorial ET chain was proposed as follows (Fig. 4C): the photoexcited P_{700} transfers the electrons to PQQ, and the electrons trapped by the relay are then transferred to the ITO electrode. Through the biocatalyzed oxidation of glucose, the $FADH_2$ cofactor in GOx is reduced. At this point, the osmium redox polymer takes the electrons from the reduced $FADH_2$—regenerating the biocatalyst—and transfers them to P_{700}^{+}, generating the PSI photosystem. This process generates increased photocurrent as the glucose concentration increases (Fig. 4D), resulting in an estimated overall light-to-electrical power conversion efficiency of 0.12%.

Furthermore, the *I-V* curve depicted in Fig. 4E shows a maximum photocurrent at −0.05 V corresponding to 450 nA. Biasing the electrode to lower potentials will, by consequence, decrease the driving force transferring the electrons from PQQ to the ITO electrode, which results in a decrease in the photocurrent. The same behavior is observed when biasing the electrode to more positive potentials, where it is seen a slight decrease in the photocurrent, a fact that is attributed to the partial oxidation of the osmium polymer. Such bioinspired system provided the basic principles for the construction of photo-biofuel cells and can be amplified to the photonic wiring of enzymes using other photoactive nanomaterials.

Not only PSI is part of the biological photosynthetic apparatus, but also photosystem II (PSII), which is able to photocatalyze water oxidation to release H^+, O_2, and electrons. For the latter, a number of biomimetic and bioinspired approaches to design protein-based electrodes have been reported in the specialized literature in order to identify undesirable light-induced charge transfer pathways [63] and to achieve Z-scheme water splitting in the artificial photosynthesis field [64–67], for instance.

Another great source of bioinspiration for the design of protein-based electrodes is the bacterial electroactive biofilms, especially regarding their ability of long-range

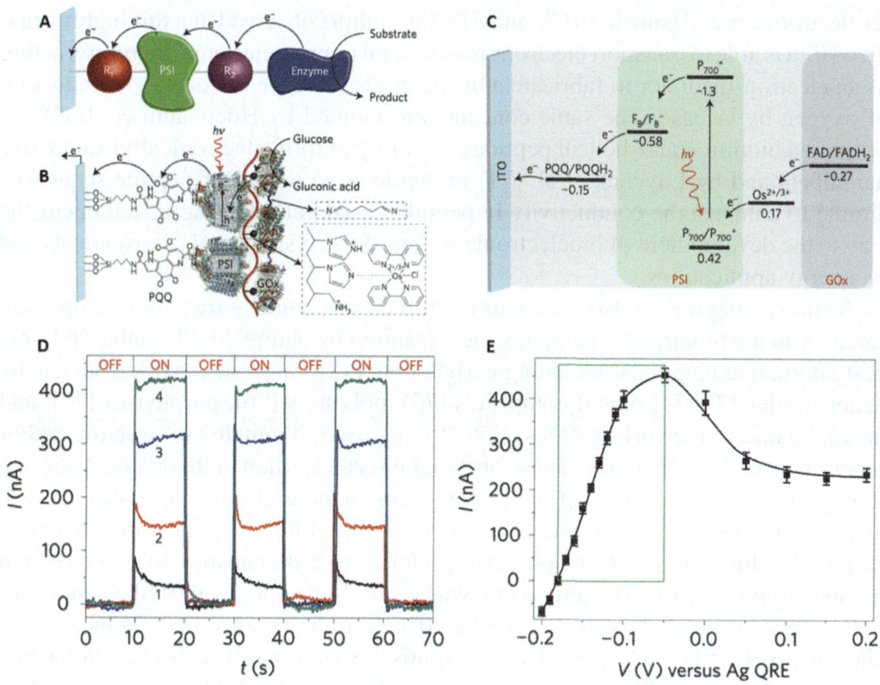

Fig. 4 Assembly of a bioinspired PSI/enzyme photoelectrochemical electrode. A General scheme for a photonically wired PSI/redox enzyme photoelectrochemical cell, in which R_1 corresponds to a relay unit wiring PSI with the electrode surface, while R_2 is a relay unit wiring the redox enzyme with PSI; **B** an organized assembly of a photonically wired PQQ/PSI/$Os^{2+/3+}$-polymer/GOx on an ITO electrode acting as a photoelectrochemical enzymatic electrode. **Energy level diagram of the redox components of the photoelectrochemical electrode. C** The organization of the energy levels of the different components allows the ET cascade yielding the photocurrent. **Photocurrent features generated by the PQQ/PSI/$Os^{2+/3+}$-polymer/GOx photo-bioelectrochemical electrode. D** ON/OFF switchable photocurrents on irradiation of the layered electrode with white-light in the presence of variable concentrations of glucose (1) 0 mM, (2) 10 mM, (3) 20 mM, (4) 30 mM. Electrode biased at 0.0 V *vs.* Ag QRE; **E** *I-V* curve corresponding to the photocurrent generated by the cell where the photoactive electrode acted as the working electrode and a Pt plat and Ag wire acted as counter and quasi-reference electrodes, respectively, at different bias potentials between the reference electrode and the working electrode, taken under white-light irradiation and with a glucose concentration of 30 mM. Reprinted by permission from Springer Nature: Nature Energy [62], copyright 2016

electron transport [68]. It is known that these biofilms are constituted by a network of conductive nanowires connecting bacteria to each other or to metal oxides in the surroundings and that the electronic conductivity is deriving from electron hopping between the redox centers. Inspired by these principles, Forge and coworkers [69] reported the design of a conductive protein bionanowires by coupling a chimeric protein resulting from the attachment of a prion domain to a rubredoxin that acts as an electron carrier. Using both CV and EIS, the authors observed that this highly organized film is able to transport electrons over several micrometers and, therefore, acting as an electron-mediator to fabricate a bioelectrode for the electrocatalytic reduction of oxygen by laccase. The same concept was adopted by Hochbaum et al. [70] to propose a biomimetic α-helical peptides assembly to form electronically conductive nanofibers and by Laycock et al. [71] in the de novo design of peptide sequences aiming to enhance the conductivity in peptide-based gels. This approach opens the way to the development of bioelectronic devices for physiological environments and bioenergy applications.

Artificial enzymes—also known as biomimetic enzymes—are another important category in the biomimetic field, in which, inspired by nature, highly stable and low-cost alternative materials, including carbon dots [72], rare earth and ferromagnetic nanoparticles [73, 74], metal complexes [75], polymers [76], porphyrins [77], and metal organic frameworks (MOFs) [78, 79], are used to imitate the structure and/or function of natural enzymes. These artificial enzymes, often called nanozymes, are projected to overcome the intrinsic drawbacks of natural enzymes, such as ease denaturation, high cost, and difficulties in recycling, finding applications in a wide range of fields, coming from biosensing and cancer diagnostics to catalysis and pollutant removal [80]. The term nanozymes, which specifically describe nanomaterials with intrinsic enzyme-like characteristics, were first coined in 2004 by Scrimin and coworkers [81] that reported the transphosphorylation activity of functionalized gold nanoparticles, followed by Gao et al. that described in 2007 the peroxidase-like property of Fe_3O_4 nanoparticles, catalyzing the oxidation of peroxidase substrates in the presence of H_2O_2 in a manner similar to that of the iron active sites within the heme group of natural enzyme horseradish peroxidase [82]. Like natural enzymes, the activity of nanozymes can be tuned by adjusting a series of parameters such as size, shape and morphology, surface coating and functionalization, composition, by the addition of activators, and by the media pH and temperature [83].

In the field of electrochemistry and, more specifically, in the bioelectrochemistry, artificial enzymes have also found important applications. For instance, not only peroxidase nanozymes have been widely employed in the construction of electrochemical affinity biosensors—which are based on registering changes in electrochemical readouts before and after specific affinity reactions—but also superoxide dismutases, hydrolases, catalases, and others [84–89]. Some works report the use of artificial peroxidase gold nanoparticles for antibiotics (kanamycin) [90] and bacteria (*Pseudomonas aeruginosa*) [91] determination in spiked acacia honey and in water, respectively. The activity was monitored by DPV and chronoamperometry, and the methods were able to detect selectively 0.73 nM of kanamycin or 60 colony-forming units per mL of *P. aeruginosa* in only 10 min. Moreover, Crespilho and coworkers [92] recently reported the use of magnetic iron oxide nanoparticles (Fe_3O_4-NPs and γ-Fe_2O_3-NPs) as redox biomimetics of complexes III and IV (cytochrome bc_1 and cytochrome oxidase, respectively) of the mitochondrial respiratory chain. The designed bioinspired system (Fig. 5A) was based on the incorporation of these nanoparticles in a mitochondrial model membrane allowing direct electron transfers between the latter and cytochrome c, acting here as an electron carrier protein,

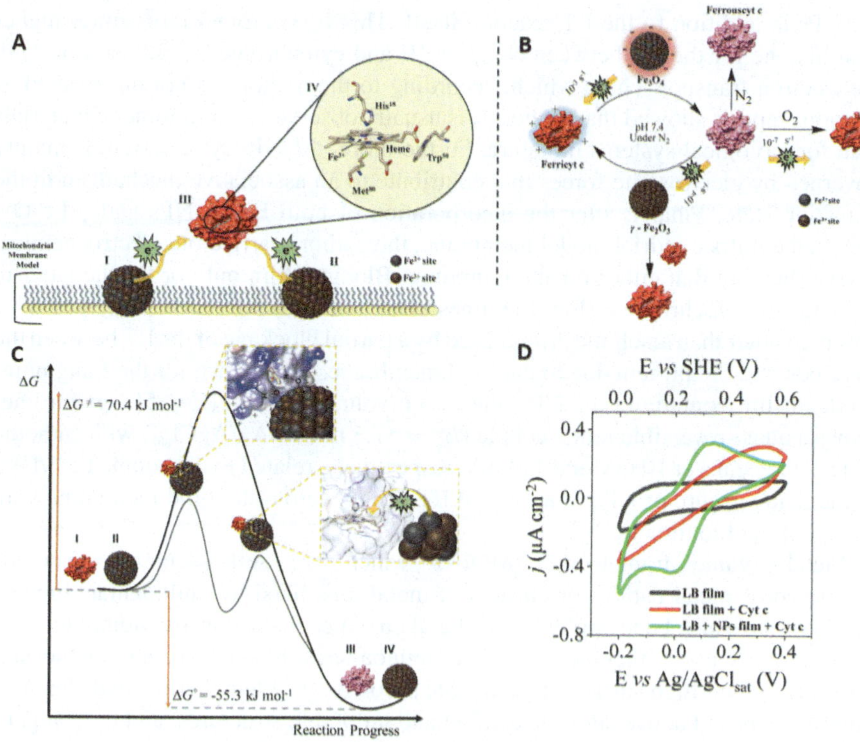

Fig. 5 Illustration of the bioinspired mitochondrial electron transfer chain. A Representation of the biomimetic model membrane with Fe_3O_4-NPs (I) and γ-Fe_2O_3 (II) incorporated in a monolayer mitochondrial cell model membrane, mimicking the redox properties of cytochrome bc_1 and cytochrome oxidase, respectively. The cytochrome c (III)—not incorporated in the membrane—has a heme group (IV) in which the iron changes between Fe^{3+} and Fe^{2+} in redox reactions. **B** Schematic illustration with calculated pseudo-first-order rate constants of the ET reactions. **C** Representation of the ferricytochrome c (I) reduction by Fe_3O_4-NPs (II), with $\Delta G^{\ddagger}$ and $\Delta G^{\circ}_{binding}$ energies, in which the first one requires Coulombic and van der Waals interactions between the species, followed by ET from the NP Fe^{2+} sites to the protein heme group, resulting in ferrouscytochrome c (III) and a partially oxidized NP (IV). **Electrochemical characterization. D** CV curves for the LB film (black), LB film + Cyt c (red) and LB + NPs film + Cyt c (green) under inert atmosphere (PBS buffer 0.1 mol L^{-1}, pH 7.4 as supporting electrolyte, scan rate of 5 mV s^{-1} at 25 °C). Reprinted from [92], copyright 2021, with permission from Elsevier

such as in the biological chain. The heterogeneous ET between nanosized iron oxides and cytochrome c was investigated in a molecular level under different conditions, in order to determine the pseudo-first-order rate constants (Fig. 5B) and the electron pathways between the inorganic biomimetic materials and the protein. The authors observed a direct and entropy-driven ET, with rate constant of 2.63 ± 0.05 L mol^{-1} at 25 °C, between the iron sites of the NPs and the cytochrome c heme group requiring an activation energy $E_a = 40.2 \pm 1.5$ kJ mol^{-1} with an overall Gibbs free energy of -55 kJ mol^{-1} (Fig. 5C), related to the binding energy between the protein and the NPs in addition to the ET reaction itself. This E_a is around three times higher than the one for the ET between complex III and cytochrome c (14.8 kJ mol^{-1}) in the electron transport chain which, according to the authors, is not unexpected as previous reports showed that higher E_a is usually observed for biomimetic materials than for biological systems in nature. Furthermore, the NP-Cyt c system is mainly governed by electrostatic forces that contribute to an associative mechanism in the transition state. Finally, after the incorporation of both Fe_3O_4-NPs and γ-Fe_2O_3-NPs in the mitochondrial model membrane, the authors carried out electrochemical assays showing that either for the Langmuir–Blodgett film only or for the latter in presence of cytochrome c (Fig. 5D, lines black and red), no redox processes were observed other than an ohmic drop caused by a partial blockage of the ET between the electrode and the protein due to the lipid membrane. In contrast, for the Langmuir–Blodgett film containing the NPs, the cyclic voltammogram (Fig. 5D, green line) shows a quasi-reversible redox couple ($E^0 = 57.5$ mV *vs* $Ag/AgCl_{sat}$) with cathodic and anodic peaks at 10 mV and 123 mV, respectively, related to the couple Fe^{3+}/Fe^{2+} of the heme group, proving that indeed ET occurs across the lipid membrane with the aid of the biomimetic NPs.

Metal–organic frameworks (MOFs)—which are self-assembled materials composed of metal cations or clusters of metal ions liked by multidentate organic ligands—have also been applied as artificial enzymes in the construction of electrochemical biosensors, therapeutics and immunoassays, bioanalysis, and as biocatalysts due to their high surface area, tunable porosity, excellent thermal and chemical stability, exposed active sites, diversified and tailorable structures, and biocompatibility [93]. These materials often share many structural and/or electronic characteristics comparable with the active sites of metalloenzymes (Fig. 6A). For instance, the metallic cations are often coordinated to the organic linkers by functionalities similar to the carboxylate, histidine, and thiolate moieties that usually form the coordination sphere in natural metalloenzymes resulting in high-spin electronic configurations, due their weak-field ligand properties [94].

Inspired by the multiple functions of MOFs in biomimetic catalysis, Ling et al. [95] proposed the design of a Pt-decorated metalloporphyrinic MOF formed by Fe porphyrin and Zr^{4+} ions that exhibited high catalase- and peroxidase-like activities, as well as high electrocatalytic activity toward H_2O_2 and O_2 reduction. The same behavior was observed by Ju and coworkers [96] when studying assemblies of Fe porphyrin on porous carbon derived from MOFs. The authors observed an excellent electrocatalytic activity of the Fe porphyrin mimicking cytochrome c oxidase toward the O_2 reduction. Furthermore, by coupling the biomimetic catalyst to a glucose oxidase enzyme, they were able to construct a biosensor for glucose detection with a linear range of 0.5–18 mM. Although most of the published works related to biomimetic MOFs is focused in heme mimicry (peroxidase activity), these materials can also mimicry the properties of phosphotriesterases (which are enzymes responsible to hydrolyze the phosphate ester bonds) as shown by Hupp et al. [97], chymotrypsin—a H-bond donating serine protease—by the incorporation of an acidic hydrogen-bond-donating squaramide moiety into a porous UiO-67 MOF derivative [98], and carbonic anhydrases, responsible for the biomineralization of CO_2 through

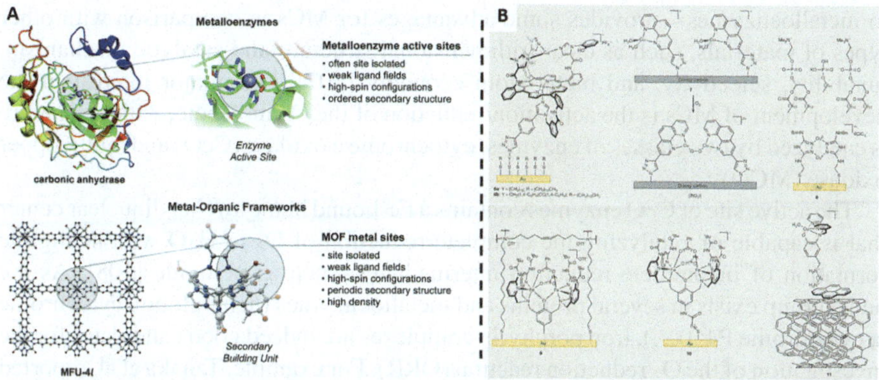

Fig. 6 Metalloenzymes and MOFs. A Properties comparison between natural metalloenzymes actives sites and metal sites in MOFs. Republished with permission of Royal Society of Chemistry, from [94], copyright 2020; permission conveyed through Copyright Clearance Center, Inc. **B** Some examples of iron and cupper-based molecular catalysts supported in metal or carbon electrodes. Cu-catalyzed alkyne-azide cycloaddition is used for the immobilization on the electrode surface, as showed for the Fe-porphyrin systems in **6** and the Cu-phenanthroline complex in **7**. If the complex has a thiol group, the binding on Au surfaces gives immobilized species as in **8** and **10**. The attachment can also be carried out by metathesis polymerization on graphite electrodes (shown in **9**) or even π-π stacking between aromatic moieties and carbon nanotubes, as in **11**. Reprinted by permission from Springer Nature: Nature Reviews Chemistry [106], copyright 2017

calcium carbonate ($CaCO_3$) crystallization [99]. Hydrogenase-like activity of MOFs was also reported; for instance, Feng and coworkers [100] were able to show that one can photocatalytically produce H_2, by combining the photosensitizing and proton reducing activity in a single Zr-porphyrin-based MOF platform. The same photocatalytic behavior was observed by Du et al. [101] when using a 2D-layered MOF with a thiolate-bridged Ni^{2+} node as a [NiFe]-hydrogenase active site biomimetic.

In addition, MOFs are suitable matrices for enzyme encapsulation due to their extremely ordered cavities/channels, providing a hydrophobic confined environment and a highly dense distribution of accessible active catalytic sites throughout their architecture [102]. With that perspective, Liu et al. [103] studied the trapping of the cytolytic peptide melittin (MLT) into porous zeolitic imidazolate framework-8 nanoparticles (ZIF-8) as a nanoplatform for delivering cytolytic peptides in cancer treatment. On the other hand, Dong and authors [104], as well as Li et al. [105], applied that same ZIF-8 framework, but now envisioning the encapsulation of glucose oxidase (GOx) for the construction of an amperometric biosensor for glucose detection. In both cases, the GOx@ZIF-8-modified electrode showed good peroxidase-like bioactivity and high electrocatalytic activity toward H_2O_2 and O_2 reduction.

Similarly to MOFs, molecular catalysts (MC) have been well-studied and applied as metalloenzymes biomimetics. Usually, a molecular catalyst is comprised by a transition metal center as the active site and specific organic ligands, producing a coordination environment that allows the regulation of the electronic structure of the active site. This configuration—which is ultimately bioinspired by the active sites in metalloenzymes—provides some advantages for MCs in comparison with other types of materials, such as clear tridimensional structure and catalytic mechanism, tunability, selectivity, and metal-atom economy [107]. One major interest in the development of MCs is the activation/reduction of the O_2 molecule, which in nature is catalyzed by two classes of enzymes: cytochrome c oxidase (C*c*O) and multicopper oxidase (MCO).

The active site of C*c*O enzymes contains a Fe-bound heme a_3/Cu_B dinuclear center that is capable of catalyzing the complete reduction of O_2 to H_2O with negligible formation of incomplete reduction intermediates, such as peroxide [108]. As the heme group exists in several proteins and metalloenzymes (hemoglobin, cytochrome c, cytochrome P450...), iron porphyrin complexes are indeed good candidates for the investigation of the O_2 reduction reaction (ORR). For example, Tanaka et al. reported a Fe^{3+}-porphyrin and Fe^{3+}-phthalocyanine complex liked through a fourfold rotaxane supported on conductive carbon black capable of efficiently reducing O_2 to H_2O with an onset potential of 0.78 V *vs* RHE in acidic conditions [109], while Anxolabéhère-Mallart and coworkers [110] investigated Fe-peroxo porphyrin complexes in the electrochemical reductive activation of O_2. On the other hand, MCO enzymes feature redox-active sites composed of two type 3 (T3), one type 2 (T2), and one type 1 (T1) Cu centers [106], operating closer to the thermodynamic potential of the O_2/H_2O couple than C*c*O. Regarding the mimicry of these redox copper centers, Cao et al. [111] and Kojima et al. [112] reported a binuclear and mononuclear Cu^{2+} complexes such as electrocatalysts for complete reduction of O_2 into H_2O. Different approaches were used to immobilize these electrocatalytic biomimetic complexes in electrodes, as shown in Fig. 6B. Finally, plenty of Co, Fe, Mn, and Cu-based molecular catalysts for electrocatalytic ORR have been studied in literature and summarized elsewhere [113–118].

Other than ORR, many different types of complexes were synthesized and studied envisioning the biomimicry of enzymes' active sites. Platinum-based complexes were proposed, as a formate dehydrogenase mimetic, to electrocatalytically convert CO_2 to HCO_2^- [119], while Ru, Ir, Mn, Co, Fe, Ni, and Cu-based molecular catalysts were employed as photosystem II biomimetics for H_2O oxidation, as summarized by Sun and Zhang [107]. Furthermore, for instance, hydrogenases and [NiFe]-carbon monoxide dehydrogenases biomimetics for H_2 evolution and conversion of CO_2 to CO, respectively, were also recently reported and summarized elsewhere [120–123].

4 Genetic Material and Live Cells

In addition to the bioinspired wiring of enzymes and proteins at electrode surfaces, the incorporation of genetic material, such as DNA, shows great potential to significantly enhance the bioelectroactivity of these biomolecules, as reported by Stieger et al. [124]. The authors applied DNA, a natural polyelectrolyte, as a building block for the construction of a photo-active 3D architecture comprising cytochrome c, as both molecular scaffold and conductive wiring network, and PSI as a photo-functional matrix element. The incorporation of DNA improved the maximum photocurrent and electrode stability, rendering this genetic material as an interesting candidate for application in photo-bioactive electrode structures.

Biological membrane channels, as controlled pathways for molecules and ions across the cell membrane, play major roles in a series of processes, such as sensing, signaling, communicating, and transporting. Due to the molecular recognition and self-assembly properties of DNA, and also inspired by the biological channels, Guo and coworkers [125] reported the usage of 3D cross-linked DNA superstructures as a biomimetic gate-like nanopore, able to manipulate the mass transportation on a subcellular scale. The authors observed that the cross-linked 3D DNA nanostructures in and out of the nanopores efficiently switch off the transverse ionic flux, changing remarkably the electrochemical resistance of the nanopores from 7.27 Ω (in the opened state) to 14.4 Ω (in the closed state). Moreover, the ON–OFF ratio between the opened and closed states approaches 10^3–10^5, which is a gating mechanism applicable to nanopores with diameters up to ca. 650 nm. Due to the high importance of biological channels and pores in the living organisms [126], the development of artificial bioinspired gate-like nanopores, such as the one reported, is fundamental for the understanding of surface-governed transport phenomena and applications on bioanalysis and smart nanofluidic devices.

Single stranded DNA (ssDNA) can also be applied as a probe for the electrochemical detection of microRNA (miRNA), according to Zhao and coworkers [127]. As miRNA is considered a biomarker in the cancer diagnosis and prognosis, the development of highly sensitive detection methods is of great clinical significance. Inspired by the biological channels, the authors reported the fabrication of a biomimetic nanochannel-ionchannel hybrid that, firstly, had ssDNA covalently immobilized on the outer surface of ionchannels through chemical coupling. This probe allowed label-free and ultrasensitive recognition of target miRNA-10b through hybridization between the latter and ssDNA. The different charge density on the outer surface and the varied effective ionchannel size after miRNA-10b recognition directly influenced the mass transport properties of the nanochannel-ionchannel. The ionic current response through the hybrid was monitored in real time using a homemade electrochemical cell, which enabled the in situ detection of miRNA with a ultralow limit of 15.4 aM.

Biomimetic and bioinspired nanomechanical DNA-based nanopores have already been designed as an approach to achieve the sequence-specific and a ligand-triggered channel opening controlled release of small molecule cargo from lipid vesicles [128]. Furthermore, Lv et al. [129] reported the construction of artificial channels on the plasma membranes of live cells, by combining DNA nanotechnology and specific chemical modifications. This system would mimicry the functions of endogenous protein channels for molecular transport. The membrane-spanning DNA nanopore, constructed using six DNA duplexes, showed a 2 nm hexagonal hollow channel, 5 nm outer diameter, and 14 nm height. To ensure a stable insertion of the biomimetic nanopore on the live cells plasma membrane, a 2-nm-high hydrophobic membrane-spanning region composed of phosphorothioate (PPT) was inserted on the nanostructure (Fig. 7A and B). After incubation, the DNA nanopores were inserted spontaneously into the plasma membrane, creating small hallow sizes of around 2 nm (Fig. 7C).

To further characterize the modified membranes, an ultralow current detection system was applied in order to examine the ion flow across the DNA nanopores inserted in the lipid bilayers in vitro (Fig. 7D). The current trace at different voltages is shown in Fig. 7E. The results showed that, comparing to the non-modified lipid bilayers, the ionic current increased steadily with higher voltages, indicating a structurally integral and stable DNA channels across the membrane over time (Fig. 7F). The authors also observed that the nanopores are able to stay on the plasma membrane of live cells for at least 20 min after insertion at 37 °C, after that they start to be internalized by the cells. Moreover, the nanopores were able to selectively transport ions in neurons cells and small molecule anticancer drugs in drug-resistant tumor cells, opening the way to applications ranging from biotechnology to biomedicine.

Real-time electrochemical monitoring of cell-released biomolecules is also quite important to understand basic cellular processes and physiological biology. Three-dimensional (3D) scaffold platforms are promising candidates to integrate electrochemical sensing, while mimicking the vivo-like cellular environment, as recently reported by Hu and coworkers [130]. In this work, the authors developed a biomimetic 3D scaffold 3-aminophenylboronic acid (APBA) functionalized graphene foam (GF) aiming the long-term cell culture and real-time electrochemical monitoring. The

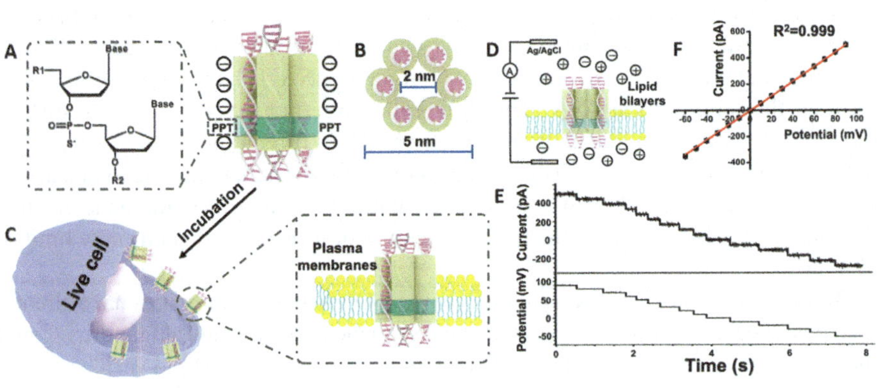

Fig. 7 Representation of the DNA-based nanopore and its insertion into the plasma membrane of live cells. A Phosphorothioate (PPT) group chemical structure modified on the DNA nanopore; **B** schematic illustration of a six-duplex -bundled DNA nanopore; **C** DNA nanopores incorporated into the plasma membrane of live cells for molecular transport. **Characterization of the DNA nanopores in vitro. D** Scheme of the ultralow current detection system with DNA nanopores incorporated in a lipid bilayer. **E** Current trace recorded at steadily increasing voltages with DNA nanopores incorporated in a lipid bilayer and **F** the corresponding current–voltage curve (solution of 1.0 mol L^{-1} KCl, 10 mM Tris, and 1.0 mM EDTA at pH 8.0). Reprinted with permission from [129], copyright 2020 American Chemical Society

biomimetic system combines both the cell-adhesive properties of APBA and the mechanical and electrochemical properties of GF. HeLa and GFP-HeLa cells grew and proliferated for a long period along the skeletons of the 3D GF, exhibiting high viability, while the release of gaseous messenger H_2S from the cells culture for different time periods was successfully monitored in real time.

5 Further Applications

Several other biomimetic and bioinspired approaches can be found in the literature in the bioelectrochemistry field. Some examples include a *Setaria viridis*-bioinspired electrode for high-performance supercapacitor application, in which the authors improved the electrochemical energy storage, the specific capacitance, and the cycling stability by mimicking the hierarchically multi-scale topography of a plant [131]; a biomimetic electroactive nanocomposite from self-assembled collagen fibrils and silver nanowires, which presented an excellent charge storage capacities and injection, as well as antimicrobial properties, while retaining the mechanical properties similar to soft tissues [132]. This approach resulted in a novel and promising bioelectrode material and electroactive tissue regenerating scaffolds. Still in the field of nanocomposites, Jiao et al. [133] recently designed a electricity-adaptive and highly-reinforced nanocellulose nanocomposite bioinspired by the ability of echinoderms and sea cucumbers to alter the stiffness of their inner dermis under certain conditions. The authors combined biosourced and sustainable wood-based cellulose nanofibrils (CNFs) with water-soluble copolymers equipped with thermo-reversible supramolecular motifs that would allow a reversible modulation of mechanical properties using low voltage direct current (DC). Applying low DC at specific voltages, they observed the dynamization and breakage of the thermo-reversible supramolecular bonds that led to a significant electrochemical softening of the nanocomposites. Moreover, these changes in the mechanical properties could be reversibly controlled by power on/power off cycles, which open new possibilities for the construction of mechanically adaptive materials triggered by readily available low DC supplies.

Lithium-ion batteries (LIBs) are a well-spread and valuable energy storage systems as they present high energy density, long cycling life, and high voltage [134]. As these systems are intrinsically depend on efficient and selective transport of lithium ions, Shi et al. [135], inspired by the ion-channels in biological systems, reported the construction of chemically modified nanoporous channels in metal–organic frameworks (MOFs) that electrostatically repelled anions while allowed the passage of lithium ions. When applying this bioinspired MOF as an electrolyte membrane, the authors observed significant enhances in lithium-ion transference number, as well as an improvement in the rate performance and in the cycling life of the LIB. Other approaches to use rechargeable lithium metal anode in LIBs were already proposed inspired by the nest of ants [136], for instance, as they are a model of efficient interconnected channels able to quickly transport air. The bioinspired ionogel electrolyte (BAIE) consisted of confining ionic liquid within interconnected SiO_2 channels to promote the dendrite-free stripping of lithium anodes with high ionic conductivity and thermal stability, as observed by the authors.

Although LIBs are important materials for energy storage, the limited availability of lithium threatens and often drives the research for new materials with similar electrochemical performance and natural abundance, such as the sodium-ion batteries (SIBs) [137]. For that, another type of material, other than graphite, needs to be used as anode, in which tin phosphide (Sn_4P_3) is a potential candidate due to its low redox potential (~0.3 V *vs* Na/Na^+) and high capacity (1132 mA h g^{-1}) [138]. However, the use of pure Sn_4P_3 is limited, as its low ionic and electronic conductivity directly affects the capacitance and cycling stability of the material. Recently, Ran and coworkers [139] reported the development of a bottlebrush-like heterostructure of Sn_4P_3 grown on carbon nanotube (Sn4P3@CNT/C) in order to improve and solve the limitations involving the direct application of this promising anode material for SIBs. To do so, firstly, the precursor SnO_2@CNT was synthesized through a hydrothermal reaction, consisting of a central CNT stem and SnO_2 nanorods of around 450 nm in length and 30 nm in diameter, resembling the bottlebrush plant (Fig. 8A). Secondly, a carbon layer was coated on the SnO_2@CNT surface also by a hydrothermal method, followed by sintering in nitrogen. The SnO_2@CNT/C still maintains the bottlebrush-like morphology, as shown in Fig. 8B. Finally, the material was hand-milled with NaH_2PO_2, heated at 280 °C, and washed with dilute HCl and water. At this point, the heterostructure Sn_4P_3@CNT/C resembles the bottlebrush fruit, in which the Sn_4P_3 nanoparticles are firmly attached to the CNT stem (Fig. 8C).

The electrochemical performance of the heterostructure was investigated, firstly concerning the discharge/charge voltage profiles at a current density of 0.2C (200 mA g^{-1}), in which the authors observed a first discharge and charge capacities of 1377 and 1173 mA h g^{-1}, respectively, corresponding to a coulombic efficiency (CE) of 85.2%. The Sn_4P_3@CNT/C CE jumps to 95% at the second cycle; it reaches to 99% within 10 cycles and keeps stable afterward still delivering 742 mA h g^{-1} after 150 cycles, as shown in Fig. 8D. Furthermore, the authors explored the electrochemical performance of the full-cell using Sn_4P_3@CNT/C as anode and $Na_3V_2(PO_4)_3$/C as the cathode (Fig. 8E). The initial charge/discharge curve profile is displayed in Fig. 8F; the obtained initial discharge capacity was of around 104 mA h g^{-1} at 0.2C with an average operating voltage of approximately 3.0 V. In terms of cycling performance, the full-cell retained 93% of its capacity after 50 cycles. In addition, the EIS results showed that the Sn_4P_3@CNT/C has a smaller impedance than the pure Sn_4P_3, and after 500 cycles, the bioinspired electrode impedance decreases, indicating an improvement in the electron/ion transport across the interface (Fig. 8G). In conclusion, the favorable electrochemical performance of the Sn_4P_3@CNT/C can be explained in terms of structure, as the CNT acts as a "stem"; providing electron expressway and mechanical stability, the Sn_4P_3 nanoparticles act as a "fructus" on the CNT surface, increasing the contact area and shortening the ion diffusion pathway, while the amorphous carbon acts as a "permeable stoma" on the tin phosphide surface, promoting the electrolyte penetration (Fig. 8H).

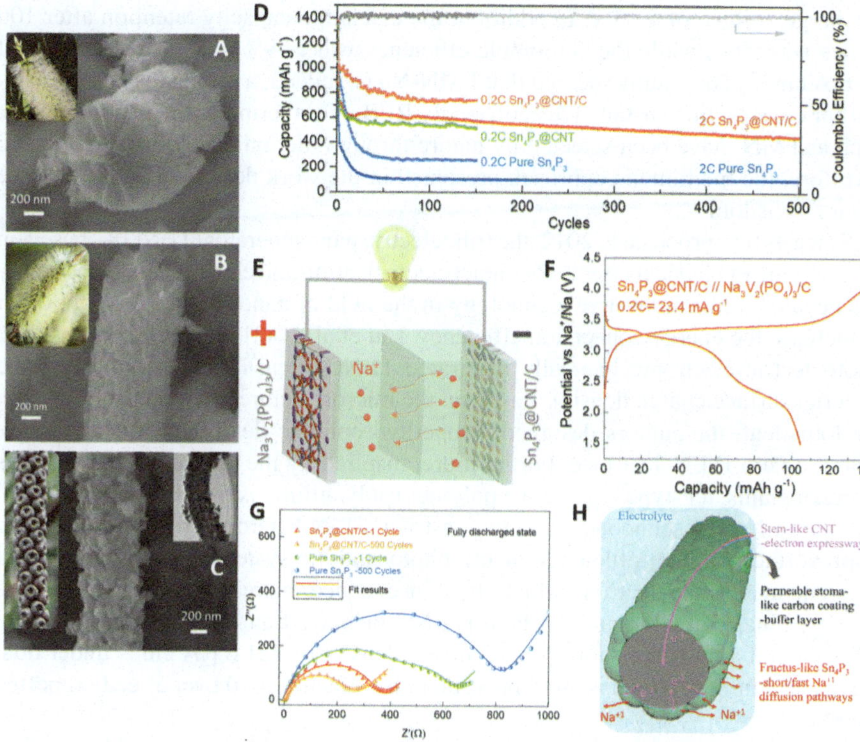

Fig. 8 SEM images of the bioinspired heterostructures with the respective bottlebrush inspiration. A SnO_2@CNT; **B** SnO_2@CNT/C; **C** Sn_4P_3@CNT/C. **Electrochemical performance of the bioinspired Sn_4P_3@CNT/C. D** Cycling and rate performance of Sn_4P_3@CNT, and pure Sn_4P_3 at a current of 0.2C (1C = 1000 mA g^{-1}) and 2.0C; **E** representation of the full-cell battery; **F** first charge/discharge curves of coin-type NVP/C//Sn_4P_3@CNT/C full-cell at 0.2C (1C = 117 mA g^{-1}); **G** EIS of pure Sn_4P_3 and Sn_4P_3@CNT/C at 1 and 500 cycles; **H** representation of the transport mechanism of Sn_4P_3@CNT/C. Reprinted with permission from [139], copyright 2020 American Chemical Society

A biomimetic approach for energy storage systems was also proposed by Orita et al. [140], but now for the construction of a aqueous redox flow battery (RFB) based on flavin mononucleotide. RFBs in general store their energy using redox active materials that are dissolved in a electrolyte—referred to as positive and negative electrolytes—which are physically separated by a membrane and circulated by pumps [141]. To apply an active material that is safe, inexpensive, and has a high-energy density, the authors chose to use a sodium salt of flavin mononucleotide (FMN-Na), known as riboflavin-5'-phosphate sodium salt, as negative electrolyte, while using potassium ferrocyanide as positive electrolyte. It is worth mentioning that flavins are cofactors of several biologically important enzymes, having one of the most versatile in vivo redox centers. The RFB showed an initial discharge capacity of 1.31 Ah l^{-1} at 10 mA cm^{-2}, an open-circuit voltage of 1.40 V and an average discharge voltage of 1.03 V. In addition, the discharge capacity retention after 100 cycles was 99%, while the coulombic efficiency was >99% at a current density of 80 mA cm^{-2}. The results showed that FMN-Na is, indeed, a promising active material for constructing sustainable ecofriendly RFBs. Furthermore, flavin compounds, such as FMN, have been selected by nature through the long evolutionary process; therefore, the biomimetic approach presented in the work derives directly from the nature's wisdom.

Since its first proposal in 2012, the triboelectric nanogenerator (TENG)—based on the coupling of physical effects of contact electrification and electrostatic induction—has been the most innovative technology in the field of nano-energy [142]. In order to increase the energy conversion efficiency, Liu et al. [143] proposed a bioinspired photoelectric-electromechanically integrated TENG (Pem-iTENG) to increase the material surface charge density. Based on the microstructure of the mastoid cilia on the lotus leaf, the authors designed a superhydrophobic flexible rod-like structure onto the Pem-iTENG surface. Furthermore, inspired by the photosynthesis process in green plants, a P-type conductive polymer (polyaniline) is matched with titanium dioxide (a N-type semiconductor) to construct a P-N heterojunction that is able to improve the quantum efficiency of the photocatalytic system, by the retention of the generated photoelectrons which, by consequence, increases the surface charge density. Finally, the bioinspired Pem-iTENG displayed an open-circuit voltage of 124.2 V and a maximal short-circuit current density of 221.6 $\mu A\ cm^{-2}$ under tidal wave and sunlight, which is an improvement by a factor of 10 over already reported TENGs.

Bioinspiration is also present in the field of electrocatalysis for renewable energies, such as the aqueous electrocatalytic reduction of CO_2 into alcohol and others hydrocarbon fuels. Recently, taking as bioinspiration the microstructure of gas-trapping cuticles of subaquatic spiders, Wakerley and coworkers [144] reported the fabrication of a superhydrophobic surface of hierarchically structured copper dendrites for CO_2 reduction. These hydrophobic gas-trapping cuticles are known as "plastron effect," which allows aquatic arachnids (such as the diving bell spider) to respire under water (Fig. 9A). The same principle was used by the authors to achieve an analogous structure, in which hierarchically structured dendritic Cu was chemically modified with a monolayer of 1-octadecanethiol to trap CO_2 gas at the electrolyte–electrode interface, as illustrated in Fig. 9B. Linear sweep voltammetry (LSV) (Fig. 9C) comparing both the hydrophobic dendrite and the equivalent wettable dendrite showed that to reach a current of -5 mA cm^{-2}, the latter required a potential of -0.68 V *vs.* RHE, whereas the hydrophobic dendrite required a more negative potential of -1.38 V *vs.* RHE. This behavior can be explained by two different contributions: firstly, a significant decrease in the electrode electrochemically active surface area and, secondly, by the lack of proton reduction activity exhibited by the hydrophobic electrode. Measurements of controlled potential electrolysis (CPE) confirmed that even at highly cathodic potentials, the bioinspired dendrite had a lowered H_2 evolution activity. For instance, at -1.6 V *vs.* RHE, the hydrophobic electrode displayed a H_2 evolution faradaic efficiency (FE) below 10%, while the wettable dendrite displayed values above 60% (Fig. 9D, E). On the other hand, the CO_2 reduction efficiency for

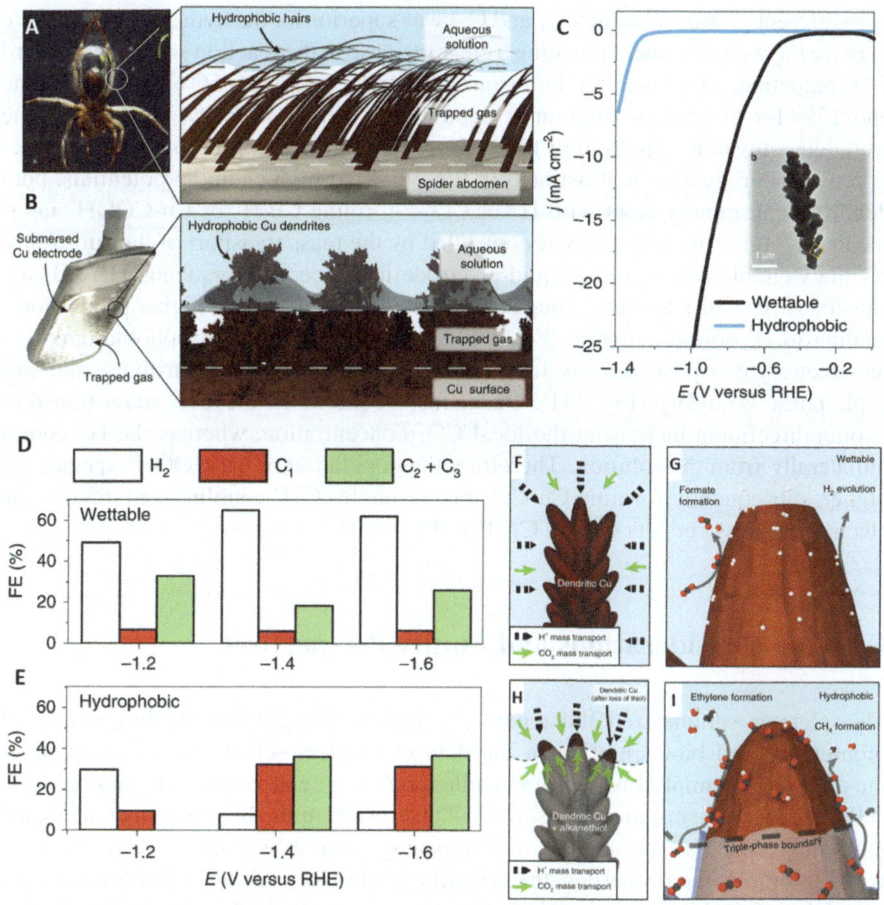

Fig. 9 Introduction to the plastron effect. A The usage of hydrophobic surfaces to trap a layer of gas between the solution-solid interface is known as plastron effect. This is a mechanism used by diving bell spiders for subaquatic breathing; **B** the same principle is applied to the hydrophobic dendritic Cu surface for aqueous CO_2 reduction. **Effect of hydrophobicity on electrocatalytic performance of the biomimetic electrode C** LSV of the wettable and hydrophobic dendrite. Inset shows a high-resolution TEM of the Cu dendrite (The electrolyte was CO_2-saturated $CsHCO_3$ 0.1 mol L^{-1}, pH 6.8, at room temperature with a scan rate of 20 mV s^{-1}); **D** CPE product FEs from the wettable and **E** from the hydrophobic dendrite at various potentials. **Proposed mechanism for CO_2 reduction. F,G** The representations of the wettable dendrite show the reactant mass transport and product formation on the electrode surface, while **H, I** depict the enhanced CO_2 mass transport from the triple-phase boundary between the electrolyte, electrode and gaseous CO_2 and the resultant formation of products on the surface of the hydrophobic dendrite. Reprinted by permission from Springer Nature: Nature Materials [144], copyright 2019

two different products (named C_1 and C_2) was superior for the hydrophobic dendrite than for the wettable one, indicating a clear increase in the reaction selectivity toward CO_2 reduction. Moreover, the bioinspired electrode attained a 56% FE for ethylene and 17% for ethanol production at neutral pH, compared to 9% and 4% on the wettable equivalent. The role of the hydrophobic surface in promoting CO_2 reduction over H^+ reduction is illustrated in Fig. 9F–I. At high cathodic potentials, both dendrites can rapidly react with H^+ or CO_2—forming Cu-H* or Cu-COOH* intermediates—thus, the selectivity is controlled by the mass transport of the substrates. As the wettable has a larger liquid-electrode interface, only aqueous H^+/CO_2 are substrates (Fig. 9F), and the groups Cu-H* are expected to be in higher proportions, leading to H_2 formation (Fig. 9G). On the contrary, in the hydrophobic dendrite, the electrolyte is pushed away from the surface, forming an electrolyte-solid–gas triple-phase boundary (Fig. 9H). Under these conditions, the CO_2 mass transport is omnidirectional increasing the local CO_2 concentration, whereas the H^+ comes unilaterally from the solution. Therefore, the population of Cu-COOH* species are higher, subsequently forming Cu-CO* that promotes C–C coupling and, finally, the electrocatalysis products C_1 and C_2 (Fig. 9I).

6 Final Considerations and Future Perspectives

This chapter summarized the recent progress and highlights the importance of biomimicry and bioinspiration in the field of bioelectrochemistry. As seen in all the discussed examples, nature is an endless rich source of inspiration, ranging from cells, microorganisms, and plants to small insects and animals. To deeply understand the complex processes occurring in biological systems, as well as, to create materials with biological-like properties, the scientific community often applies fundamental principles of the biomimicry and bioinspiration fields. Regarding this point, electrochemical techniques—such as cyclic voltammetry, electrochemical impedance spectroscopy, differential capacitance, and chronocoulometry—are quite important tools to investigate these biomimetic systems. These tools become even more useful when coupled with other techniques, such as surface-sensitive methods (spectroscopy, microscopic, etc.), especially for the characterization of biomimetic membranes and reconstituted membrane proteins, which ultimately contributes to the development of biosensors and biomedical devices. Bioinspired approaches can also be used to design protein-based electrodes and to fabricate biomimetic enzymes able to catalyze reactions in a manner similar to that of natural biological active sites. In addition, genetic material (DNA and RNA, for instance) and live cells are biomaterials that have also found application in biomimicry, including the construction of biosensors for diagnosis, gate-like nanopores for small molecules and ions delivery into cells, for enhancement in the bioelectroactivity of biomolecules, and for cell culture cultivation. Furthermore, nanostructures present in plants and small insects are a source of inspiration especially in the energy storing materials, such as batteries and nanogenerators, and bioelectrocatalysis field.

Finally, every year more and more research papers are published presenting innovative approaches for the design and study of biomimetics and/or bioinspired materials. Nevertheless, there are still several improvements needed in the field in order to obtain more complex materials to get as close as possible to the natural biological entities or even to surpass the properties nature offers. As the biological materials' structures range from the atomic to the macroscale, and considering that property is intrinsically dependent on structure, sophisticated and highly accurate characterization techniques able to access each of these length scales still need improvements. It is also needed to take in consideration that most of the biological and bioinspired/biomimetic materials are quite delicate in a manner that they can be easily damaged by, for instance, electron beams and/or vacuum conditions. Furthermore, the production of these materials in an industrial level is still limited, as they are usually fabricated and evaluated in a small quantity within the laboratory. These challenges are exciting opportunities to invest even more in the development of biomimetics that will allow, for certain, a more cooperative evolution and an environmental-friendly technological development.

References

1. Naik, R.R., Singamaneni, S.: Introduction: bioinspired and biomimetic materials. Chem. Rev. **117**, 12581–12583 (2017). https://doi.org/10.1021/acs.chemrev.7b00552
2. Suresh Kumar, N., Padma Suvarna, R., Chandra Babu Naidu, K., Banerjee, P., Ratnamala, A., Manjunatha, H.: A review on biological and biomimetic materials and their applications. Appl. Phys. A Mater. Sci. Process. **126** (2020). https://doi.org/10.1007/s00339-020-03633-z
3. Wegst, U.G.K., Bai, H., Saiz, E., Tomsia, A.P., Ritchie, R.O.: Bioinspired structural materials. Nat. Mater. **14**, 23–36 (2015). https://doi.org/10.1038/nmat4089
4. Vullev, V.I.: From biomimesis to bioinspiration: what's the benefit for solar energy conversion applications? J. Phys. Chem. Lett. **2**, 503–508 (2011). https://doi.org/10.1021/jz1016069
5. Naleway, S.E., Porter, M.M., McKittrick, J., Meyers, M.A.: structural design elements in biological materials: application to bioinspiration. Adv. Mater. **27**, 5455–5476 (2015). https://doi.org/10.1002/adma.201502403
6. Wang, Y., Naleway, S.E., Wang, B.: Biological and bioinspired materials: structure leading to functional and mechanical performance. Bioact. Mater. **5**, 745–757 (2020). https://doi.org/10.1016/j.bioactmat.2020.06.003
7. Sharma, S., Sarkar, P.: Biomimicry: exploring research, challenges, gaps, and tools. In: Chakrabarti, A. (eds.) Research into Design for a Connected World. Smart Innovation, Systems and Technologies, vol. 134, pp 87–97. Springer, Singapore (2019)
8. Swiegers, G.F.: Bioinspiration and Biomimicry in Chemistry: Reverse-Engineering Nature. Wiley, Hoboken, NJ (2012)
9. Thanigaiarasu, P.: Biomimetics in the design of medical devices. In: Sahnmugam, P.S.T., Chokkalingam, L., Bakthavachalam, P. (eds.) Trends in Development of Medical Devices, 1st edn. Elsevier, London (2020)
10. Ricke, J.: Biomimetic chemistry at interfaces. In: Ball, V. (ed.) Self-Assembly Processes at Interfaces: Multiscale Phenomena, 1st edn. Elsevier, London (2017)
11. Martín-Palma, R.J., Lakhtakia, A.: Progress on bioinspired, biomimetic, and bioreplication routes to harvest solar energy. Appl. Phys. Rev. **4** (2017). https://doi.org/10.1063/1.4981792

12. Yin, X., Müller, R.: Integration of deep learning and soft robotics for a biomimetic approach to nonlinear sensing. Nat. Mach. Intell. **3**, 507–512 (2021). https://doi.org/10.1038/s42256-021-00330-1
13. Pawlyn, M.: Biomimicry in Architecture, 2nd edn. RIBA Publishing, Newcastle (2016)
14. Park, E.-Y.: Body painting type analysis based on biomimicry camouflage. Int. J. Archit. Arts. Appl. **6**, 1–11 (2020). https://doi.org/10.11648/j.ijaaa.20200601.11
15. Lepora, N.F., Verschure, P., Prescott, T.J.: The state of the art in biomimetics. Bioinspiration and Biomimetics **8** (2013). https://doi.org/10.1088/1748-3182/8/1/013001
16. de Souza, J.C.P., Macedo, L.J.A., Hassan, A., Sedenho, G.C., Modenez, I.A., Crespilho, F.N.: In situ and operando techniques for investigating electron transfer in biological systems. ChemElectroChem **8**, 431–446 (2021). https://doi.org/10.1002/celc.202001327
17. Nieciecka, D., Królikowska, A., Krysinski, P.: Probing the interactions of mitoxantrone with biomimetic membranes with electrochemical and spectroscopic techniques. Electrochim. Acta **165**, 430–442 (2015). https://doi.org/10.1016/j.electacta.2015.02.223
18. Guidelli, R.: Bioelectrochemistry of Biomembranes and Biomimetic Membranes, 1st edn. Wiley, Hoboken, NJ (2017)
19. Bondar, A.N., Lemieux, M.J.: Reactions at biomembrane interfaces. Chem. Rev. **119**, 6162–6183 (2019). https://doi.org/10.1021/acs.chemrev.8b00596
20. Lipkowski, J.: Biomimetics: a new research opportunity for surface electrochemistry. J. Solid State Electrochem. **24**, 2121–2123 (2020). https://doi.org/10.1007/s10008-020-04627-w
21. Luchini, A., Vitiello, G.: Mimicking the mammalian plasma membrane: an overview of lipid membrane models for biophysical studies. Biomimetics **6**, 1–18 (2021). https://doi.org/10.3390/biomimetics6010003
22. Su, Z.F., Leitch, J.J., Lipkowski, J.: Electrode-supported biomimetic membranes: an electrochemical and surface science approach for characterizing biological cell membranes. Curr. Opin. Electrochem. **12**, 60–72 (2018). https://doi.org/10.1016/j.coelec.2018.05.020
23. Blank, M., Miller, I.R.: Transport of ions across lipid monolayers. I. The structure of decylammonium monolayers at the polarized mercury-water interface. J. Colloid Interface Sci. **26**, 26–33 (1968). https://doi.org/10.1016/0021-9797(68)90267-1
24. Sanver, D., Murray, B.S., Sadeghpour, A., Rappolt, M., Nelson, A.L.: Experimental modeling of flavonoid-biomembrane interactions. Langmuir **32**, 13234–13243 (2016). https://doi.org/10.1021/acs.langmuir.6b02219
25. Rashid, A., Vakurov, A., Mohamadi, S., Sanver, D., Nelson, A.: Substituents modulate biphenyl penetration into lipid membranes. Biochim. Biophys. Acta – Biomembr. **1859**, 712–721 (2017). https://doi.org/10.1016/j.bbamem.2017.01.023
26. Vakurov, A., Drummond-Brydson, R., Ugwumsinachi, O., Nelson, A.: Significance of particle size and charge capacity in TiO_2 nanoparticle-lipid interactions. J. Colloid Interface Sci. **473**, 75–83 (2016). https://doi.org/10.1016/j.jcis.2016.03.045
27. Levine, Z.A., Denardis, N.I., Vernier, P.T.: Phospholipid and hydrocarbon interactions with a charged electrode interface. Langmuir **32**, 2808–2819 (2016). https://doi.org/10.1021/acs.langmuir.5b04090
28. Pawłowski, J., Juhaniewicz, J., Güzeloǫlu, A., Sek, S.: Mechanism of lipid vesicles spreading and bilayer formation on a Au(111) surface. Langmuir **31**, 11012–11019 (2015). https://doi.org/10.1021/acs.langmuir.5b01331
29. Pawłowski, J., Dziubak, D., Sęk, S.: Potential-driven changes in hydration of chitosan-derived molecular films on gold electrodes. Electrochim. Acta **319**, 606–614 (2019). https://doi.org/10.1016/j.electacta.2019.07.021
30. Prieto, F., Rueda, M., Naitlho, N., Vázquez-González, M., González-Rodríguez, M.L., Rabasco, A.M.: Electrochemical characterization of a mixed lipid monolayer supported on Au(111) electrodes with implications for doxorubicin delivery. J. Electroanal. Chem. **815**, 246–254 (2018). https://doi.org/10.1016/j.jelechem.2018.02.056
31. Konarzewska, D., Juhaniewicz, J., Güzeloğlu, A., Sęk, S.: Characterization of planar biomimetic lipid films composed of phosphatidylethanolamines and phosphatidylglycerols from Escherichia coli. Biochim. Biophys. Acta – Biomembr. **1859**, 475–483 (2017). https://doi.org/10.1016/j.bbamem.2017.01.010

32. Juhaniewicz-Dębińska, J., Tymecka, D., Sęk, S.: Lipopeptide-induced changes in permeability of solid supported bilayers composed of bacterial membrane lipids. J. Electroanal. Chem. **812**, 227–234 (2018). https://doi.org/10.1016/j.jelechem.2017.12.065
33. Juhaniewicz, J., Szyk-Warszyńska, L., Warszyński, P., Sek, S.: Interaction of cecropin B with zwitterionic and negatively charged lipid bilayers immobilized at gold electrode surface. Electrochim. Acta **204**, 206–217 (2016). https://doi.org/10.1016/j.electacta.2016.04.080
34. Pieta, P., Majewska, M., Su, Z., Grossutti, M., Wladyka, B., Piejko, M., Lipkowski, J., Mak, P.: Physicochemical studies on orientation and conformation of a new bacteriocin BacSp222 in a planar phospholipid bilayer. Langmuir **32**, 5653–5662 (2016). https://doi.org/10.1021/acs.langmuir.5b04741
35. Su, H., Liu, H.Y., Pappa, A.M., Hidalgo, T.C., Cavassin, P., Inal, S., Owens, R.M., Daniel, S.: Facile generation of biomimetic-supported lipid bilayers on conducting polymer surfaces for membrane biosensing. ACS Appl. Mater. Interfaces **11**, 43799–43810 (2019). https://doi.org/10.1021/acsami.9b10303
36. Alvarez-Malmagro, J., Jablonowska, E., Nazaruk, E., Szwedziak, P., Bilewicz, R.: How do lipid nanocarriers—cubosomes affect electrochemical properties of DMPC bilayers deposited on gold (111) electrodes? Bioelectrochemistry **134** (2020). https://doi.org/10.1016/j.bioelechem.2020.107516
37. Gao, T., Liu, F., Yang, D., Yu, Y., Wang, Z., Li, G.: Assembly of selective biomimetic surface on an electrode surface: a design of nano-bio interface for biosensing. Anal. Chem. **87**, 5683–5689 (2015). https://doi.org/10.1021/acs.analchem.5b00816
38. Madrid, E., Horswell, S.L.: The electrochemical phase behaviour of chemically asymmetric lipid bilayers supported at Au(111) electrodes. J. Electroanal. Chem. **819**, 338–346 (2018). https://doi.org/10.1016/j.jelechem.2017.11.006
39. Andersson, J., Köper, I.: Tethered and polymer supported bilayer lipid membranes: structure and function. Membranes (Basel) **6** (2016). https://doi.org/10.3390/membranes6020030
40. Wiebalck, S., Kozuch, J., Forbrig, E., Tzschucke, C.C., Jeuken, L.J.C., Hildebrandt, P.: Monitoring the transmembrane proton gradient generated by cytochrome bo_3 in tethered bilayer lipid membranes using SEIRA spectroscopy. J. Phys. Chem. B **120**, 2249–2256 (2016). https://doi.org/10.1021/acs.jpcb.6b01435
41. Andersson, J., Fuller, M.A., Wood, K., Holt, S.A., Köper, I.: A tethered bilayer lipid membrane that mimics microbial membranes. Phys. Chem. Chem. Phys. **20**, 12958–12969 (2018). https://doi.org/10.1039/c8cp01346b
42. Zhou, W., Burke, P.J.: Versatile bottom-up synthesis of tethered bilayer lipid membranes on nanoelectronic biosensor devices. ACS Appl. Mater. Interfaces **9**, 14618–14632 (2017). https://doi.org/10.1021/acsami.7b00268
43. Su, Z., Juhaniewicz-Debinska, J., Sek, S., Lipkowski, J.: Water structure in the submembrane region of a floating lipid bilayer: the effect of an ion channel formation and the channel blocker. Langmuir **36**, 409–418 (2020). https://doi.org/10.1021/acs.langmuir.9b03271
44. Abbasi, F., Leitch, J.J., Su, Z.F., Szymanski, G., Lipkowski, J.: Direct visualization of alamethicin ion pores formed in a floating phospholipid membrane supported on a gold electrode surface. Electrochim. Acta **267**, 195–205 (2018). https://doi.org/10.1016/j.electacta.2018.02.057
45. Abbasi, F., Su, Z.F., Alvarez-Malmagro, J., Leitch, J.J., Lipkowski, J.: Effects of amiloride, an ion channel blocker, on alamethicin pore formation in negatively charged, gold-supported, phospholipid bilayers: a molecular view. Langmuir **35**, 5060–5068 (2019). https://doi.org/10.1021/acs.langmuir.9b00187
46. Ryu, H., Fuwad, A., Yoon, S., Jang, H., Lee, J.C., Kim, S.M., Jeon, T.J.: Biomimetic membranes with transmembrane proteins: state-of-the-art in transmembrane protein applications. Int. J. Mol. Sci. **20** (2019). https://doi.org/10.3390/ijms20061437
47. Kim, Y.H., Hang, L., Cifelli, J.L., Sept, D., Mayer, M., Yang, J.: Frequency-based analysis of gramicidin A nanopores enabling detection of small molecules with picomolar sensitivity. Anal. Chem. **90**, 1635–1642 (2018). https://doi.org/10.1021/acs.analchem.7b02961

48. Park, J., Lim, M.C., Ryu, H., Shim, J., Kim, S.M., Kim, Y.R., Jeon, T.J.: Nanopore based detection of: Bacillus thuringiensis HD-73 spores using aptamers and versatile DNA hairpins. Nanoscale **10**, 11955–11961 (2018). https://doi.org/10.1039/c8nr03168a
49. Zieleniecki, J.L., Nagarajan, Y., Waters, S., Rongala, J., Thompson, V., Hrmova, M., Köper, I.: Cell-free synthesis of a functional membrane transporter into a tethered bilayer lipid membrane. Langmuir **32**, 2445–2449 (2016). https://doi.org/10.1021/acs.langmuir.5b04059
50. Steininger, C., Reiner-Rozman, C., Schwaighofer, A., Knoll, W., Naumann, R.L.C.: Kinetics of cytochrome c oxidase from R. sphaeroides initiated by direct electron transfer followed by tr-SEIRAS. Bioelectrochemistry **112**, 1–8 (2016). https://doi.org/10.1016/j.bioelechem.2016.06.005
51. Karaballi, R.A., Merchant, S., Power, S.R., Brosseau, C.L.: Electrochemical surface-enhanced Raman spectroscopy (EC-SERS) study of the interaction between protein aggregates and biomimetic membranes. Phys. Chem. Chem. Phys. **20**, 4513–4526 (2018). https://doi.org/10.1039/c7cp06838g
52. Gutiérrez-Sanz, Ó., Tapia, C., Marques, M.C., Zacarias, S., Vélez, M., Pereira, I.A.C., De Lacey, A.L.: Induction of a proton gradient across a gold-supported biomimetic membrane by electroenzymatic H_2 oxidation. Angew. Chemie. Int. Ed. **54**, 2684–2687 (2015). https://doi.org/10.1002/anie.201411182
53. Gutiérrez-Sanz, Ó., Natale, P., Márquez, I., Marques, M.C., Zacarias, S., Pita, M., Pereira, I.A.C., López-Montero, I., De Lacey, A.L., Vélez, M.: H_2-fueled ATP synthesis on an electrode: mimicking cellular respiration. Angew. Chemie. Int. Ed. **55**, 6216–6220 (2016). https://doi.org/10.1002/anie.201600752
54. García-Molina, G., Natale, P., Valenzuela, L., Alvarez-Malmagro, J., Gutiérrez-Sánchez, C., Iglesias-Juez, A., López-Montero, I., Vélez, M., Pita, M., De Lacey, A.L. Potentiometric detection of ATP based on the transmembrane proton gradient generated by ATPase reconstituted on a gold electrode. Bioelectrochemistry **133** (2020). https://doi.org/10.1016/j.bioelechem.2020.107490
55. Niroomand, H., Pamu, R., Mukherjee, D., Khomami, B.: Microenvironment alterations enhance photocurrents from photosystem i confined in supported lipid bilayers. J. Mater. Chem. A **6**, 12281–12290 (2018). https://doi.org/10.1039/c8ta00898a
56. Heath, G.R., Li, M., Rong, H., Radu, V., Frielingsdorf, S., Lenz, O., Butt, J.N., Jeuken, L.J.C.: Multilayered lipid membrane stacks for biocatalysis using membrane enzymes. Adv. Funct. Mater. **27** (2017). https://doi.org/10.1002/adfm.201606265
57. Mazurenko, I., Hitaishi, V.P., Lojou, E.: Recent advances in surface chemistry of electrodes to promote direct enzymatic bioelectrocatalysis. Curr. Opin. Electrochem. **19**, 113–121 (2020). https://doi.org/10.1016/j.coelec.2019.11.004
58. Milton, R.D., Minteer, S.D.: Direct enzymatic bioelectrocatalysis: differentiating between myth and reality. J. R. Soc. Interface **14** (2017). https://doi.org/10.1098/rsif.2017.0253
59. Saboe, P.O., Conte, E., Farell, M., Bazan, G.C., Kumar, M.: Biomimetic and bioinspired approaches for wiring enzymes to electrode interfaces. Energy Environ. Sci. **10**, 14–42 (2017). https://doi.org/10.1039/c6ee02801b
60. Pereira, A.R., Sedenho, G.C., de Souza, J.C.P., Crespilho, F.N.: Advances in enzyme bioelectrochemistry. An. Acad. Bras. Cienc. **90**, 825–857 (2018). https://doi.org/10.1590/0001-3765201820170514
61. Chowdhury, R., Stromer, B., Pokharel, B., Kumar, C.V.: Control of enzyme-solid interactions via chemical modification. Langmuir **28**, 11881–11889 (2012). https://doi.org/10.1021/la3022003
62. Efrati, A., Lu, C.-H., Michaeli, D., Nechushtai, R., Alsaoub, S., Schuhmann, W., Willner, I.: Assembly of photo-bioelectrochemical cells using photosystem I-functionalized electrodes. Nat. Energy **1**, 1–8 (2016). https://doi.org/10.1038/nenergy.2015.21
63. Zhang, J.Z., Sokol, K.P., Paul, N., Romero, E., Van Grondelle, R., Reisner, E.: Competing charge transfer pathways at the photosystem II-electrode interface. Nat. Chem. Biol. **12**, 1046–1052 (2016). https://doi.org/10.1038/nchembio.2192

64. Herzallh, N.S., Cohen, Y., Mukha, D., Neumann, E., Michaeli, D., Nechushtai, R., Yehezkeli, O.: Photosynthesis Z-Scheme biomimicry: photosystem I/$BiVO_4$ photo-bioelectrochemical cell for donor-free bias-free electrical power generation. Biosens. Bioelectron. **168**, 112517 (2020). https://doi.org/10.1016/j.bios.2020.112517
65. Yu, W., Zhang, S., Chen, J., Xia, P., Richter, M.H., Chen, L., Xu, W., Jin, J., Chen, S., Peng, T.: Biomimetic Z-scheme photocatalyst with a tandem solid-state electron flow catalyzing H_2 evolution. J. Mater. Chem. A **6**, 15668–15674 (2018). https://doi.org/10.1039/c8ta02922a
66. Li, Z., Wang, W., Ding, C., Wang, Z., Liao, S., Li, C.: Biomimetic electron transport via multi-redox shuttles from photosystem II to a photoelectrochemical cell for solar water splitting. Energy Environ. Sci. **10**, 765–771 (2017). https://doi.org/10.1039/c6ee03401b
67. Sokol, K.P., Robinson, W.E., Warnan, J., Kornienko, N., Nowaczyk, M.M., Ruff, A., Zhang, J.Z., Reisner, E.: Bias-free photoelectrochemical water splitting with photosystem II on a dye-sensitized photoanode wired to hydrogenase. Nat Energy **3**, 944–951 (2018). https://doi.org/10.1038/s41560-018-0232-y
68. Ing, N.L., Nusca, T.D., Hochbaum, A.I.: Geobacter sulfurreducens pili support ohmic electronic conduction in aqueous solution. Phys. Chem. Chem. Phys. **19**, 21791–21799 (2017). https://doi.org/10.1039/c7cp03651e
69. Altamura, L., Horvath, C., Rengaraj, S., Rongier, A., Elouarzaki, K., Gondran, C., Maçon, A.L.B., Vendrely, C., Bouchiat, V., Fontecave, M., Mariolle, D., Rannou, P., Le Goff, A., Duraffourg, N., Holzinger, M., Forge, V.: A synthetic redox biofilm made from metalloprotein-prion domain chimera nanowires. Nat. Chem. **9**, 157–163 (2017). https://doi.org/10.1038/NCHEM.2616
70. Ing, N.L., Spencer, R.K., Luong, S.H., Nguyen, H.D., Hochbaum, A.I.: Electronic conductivity in biomimetic α-helical peptide nanofibers and gels. ACS Nano **12**, 2652–2661 (2018). https://doi.org/10.1021/acsnano.7b08756
71. Creasey, R.C.G., Mostert, A.B., Solemanifar, A., Nguyen, T.A.H., Virdis, B., Freguia, S., Laycock, B.: Biomimetic peptide nanowires designed for conductivity. ACS Omega **4**, 1748–1756 (2019). https://doi.org/10.1021/acsomega.8b02231
72. Vázquez-González, M., Liao, W.C., Cazelles, R., Wang, S., Yu, X., Gutkin, V., Willner, I.: Mimicking horseradish peroxidase functions using Cu^{2+}-modified carbon nitride nanoparticles or Cu^{2+}-modified carbon dots as heterogeneous catalysts. ACS Nano **11**, 3247–3253 (2017). https://doi.org/10.1021/acsnano.7b00352
73. Kluenker, M., Nawaz Tahir, M., Ragg, R., Korschelt, K., Simon, P., Gorelik, T.E., Barton, B., Shylin, S.I., Panthöfer, M., Herzberger, J., Frey, H., Ksenofontov, V., Möller, A., Kolb, U., Grin, J., Tremel, W.: Pd@Fe_2O_3 superparticles with enhanced peroxidase activity by solution phase epitaxial growth. Chem. Mater. **29**, 1134–1146 (2017). https://doi.org/10.1021/acs.chemmater.6b04283
74. Sang, J., Wu, R., Guo, P., Du, J., Xu, S., Wang, J.: Affinity-tuned peroxidase-like activity of hydrogel-supported Fe_3O_4 nanozyme through alteration of crosslinking concentration. J. Appl. Polym. Sci. **133**, 1–10 (2016). https://doi.org/10.1002/app.43065
75. Bhunia, S., Rana, A., Roy, P., Martin, D.J., Pegis, M.L., Roy, B., Dey, A.: Rational design of mononuclear iron porphyrins for facile and selective $4e^-/4H^+$ O_2 reduction: activation of O-O bond by 2nd sphere hydrogen bonding. J. Am. Chem. Soc. **140**, 9444–9457 (2018). https://doi.org/10.1021/jacs.8b02983
76. Li, M., Zhao, M., Cao, X., Zhao, M.: Smart polymers for biomimetic catalysis and enzyme inhibition. In: Li, S., Lieberzeit, P.A., Piletsky, S.A., Turner, A.P.F. (eds.) Smart Polymer Catalysis and Tunable Catalysis, 1st edn. Elsevier, Amsterdam (2019)
77. Hötger, D., Etzkorn, M., Morchutt, C., Wurster, B., Dreiser, J., Stepanow, S., Grumelli, D., Gutzler, R., Kern, K.: Stability of metallo-porphyrin networks under oxygen reduction and evolution conditions in alkaline media. Phys. Chem. Chem. Phys. **21**, 2587–2594 (2019). https://doi.org/10.1039/c8cp07463a
78. Lin, T., Qin, Y., Huang, Y., Yang, R., Hou, L., Ye, F., Zhao, S.: A label-free fluorescence assay for hydrogen peroxide and glucose based on the bifunctional MIL-53(Fe) nanozyme. Chem. Commun. **54**, 1762–1765 (2018). https://doi.org/10.1039/c7cc09819g

79. Ortiz-Gómez, I., Salinas-Castillo, A., García, A.G., Álvarez-Bermejo, J.A., de Orbe-Payá, I., Rodríguez-Diéguez, A., Capitán-Vallvey, L.F.: Microfluidic paper-based device for colorimetric determination of glucose based on a metal-organic framework acting as peroxidase mimetic. Microchim. Acta **185** (2018). https://doi.org/10.1007/s00604-017-2575-7
80. Wei, H., Wang, E.: Nanomaterials with enzyme-like characteristics (nanozymes): next-generation artificial enzymes. Chem. Soc. Rev. **42**, 6060–6093 (2013). https://doi.org/10.1039/c3cs35486e
81. Manea, F., Houillon, F.B., Pasquato, L., Scrimin, P.: Nanozymes: gold-nanoparticle-based transphosphorylation catalysts. Angew. Chemie. **116**, 6291–6295 (2004). https://doi.org/10.1002/ange.200460649
82. Gao, L., Zhuang, J., Nie, L., Zhang, J., Zhang, Y., Gu, N., Wang, T., Feng, J., Yang, D., Perrett, S., Yan, X.: Intrinsic peroxidase-like activity of ferromagnetic nanoparticles. Nat. Nanotechnol. **2**, 577 (2007). https://doi.org/10.1038/nnano.2007.260
83. Liang, M., Yan, X.: Nanozymes: from new concepts, mechanisms, and standards to applications. Acc. Chem. Res. **52**, 2190–2200 (2019). https://doi.org/10.1021/acs.accounts.9b00140
84. Campuzano, S., Pedrero, M., Yáñez-Sedeño, P., Pingarrón, J.M.: Nanozymes in electrochemical affinity biosensing. Microchim. Acta **187** (2020). https://doi.org/10.1007/s00604-020-04390-9
85. Peng, F., Xu, T., Wu, F., Ma, C., Liu, Y., Li, J., Zhao, B., Mao, C.: Novel biomimetic enzyme for sensitive detection of superoxide anions. Talanta **174**, 82–91 (2017). https://doi.org/10.1016/j.talanta.2017.05.028
86. Chatterjee, B., Das, S.J., Anand, A., Sharma, T.K.: Nanozymes and aptamer-based biosensing. Mater. Sci. Energy Technol. **3**, 127–135 (2020). https://doi.org/10.1016/j.mset.2019.08.007
87. Komkova, M.A., Andreeva, K.D., Zarochintsev, A.A., Karyakin, A.A.: Nanozymes "artificial peroxidase": enzyme oxidase mixtures for single-step fabrication of advanced electrochemical biosensors. ChemElectroChem **8**, 1117–1122 (2021). https://doi.org/10.1002/celc.202100275
88. Niu, X., Cheng, N., Ruan, X., Du, D., Lin, Y.: Review—Nanozyme-based immunosensors and immunoassays: recent developments and future trends. J. Electrochem. Soc. **167**, 037508 (2020). https://doi.org/10.1149/2.0082003jes
89. Mahmudunnabi, R.G., Farhana, F.Z., Kashaninejad, N., Firoz, S.H., Shim, Y.B., Shiddiky, M.J.A.: Nanozyme-based electrochemical biosensors for disease biomarker detection. Analyst **145**, 4398–4420 (2020). https://doi.org/10.1039/d0an00558d
90. Wang, C., Liu, C., Luo, J., Tian, Y., Zhou, N.: Direct electrochemical detection of kanamycin based on peroxidase-like activity of gold nanoparticles. Anal. Chim. Acta **936**, 75–82 (2016). https://doi.org/10.1016/j.aca.2016.07.013
91. Das, R., Dhiman, A., Kapil, A., Bansal, V., Sharma, T.K.: Aptamer-mediated colorimetric and electrochemical detection of Pseudomonas aeruginosa utilizing peroxidase-mimic activity of gold NanoZyme. Anal. Bioanal. Chem. **411**, 1229–1238 (2019). https://doi.org/10.1007/s00216-018-1555-z
92. Modenez, I.A., Macedo, L.J.A., Melo, A.F.A.A., Pereira, A.R., Oliveira, O.N., Crespilho, F.N.: Nanosized non-proteinaceous complexes III and IV mimicking electron transfer of mitochondrial respiratory chain. J. Colloid Interface Sci. **599**, 198–206 (2021). https://doi.org/10.1016/j.jcis.2021.04.072
93. Begum, S., Hassan, Z., Bräse, S., Wöll, C., Tsotsalas, M.: Metal-organic framework-templated biomaterials: recent progress in synthesis, functionalization, and applications. Acc. Chem. Res. **52**, 1598–1610 (2019). https://doi.org/10.1021/acs.accounts.9b00039
94. Bour, J.R., Wright, A.M., He, X., Dincă, M.: Bioinspired chemistry at MOF secondary building units. Chem. Sci. **11**, 1728–1737 (2020). https://doi.org/10.1039/c9sc06418d
95. Ling, P., Cheng, S., Chen, N., Qian, C., Gao, F.: Nanozyme-modified metal-organic frameworks with multienzymes activity as biomimetic catalysts and electrocatalytic interfaces. ACS Appl. Mater. Interfaces **12**, 17185–17192 (2020). https://doi.org/10.1021/acsami.9b23147
96. Ling, P., Hao, Q., Lei, J., Ju, H.: Porphyrin functionalized porous carbon derived from metal-organic framework as a biomimetic catalyst for electrochemical biosensing. J. Mater. Chem. B **3**, 1335–1341 (2015). https://doi.org/10.1039/c4tb01620c

97. Katz, M.J., Mondloch, J.E., Totten, R.K., Park, J.K., Nguyen, S.T., Farha, O.K., Hupp, J.T.: Simple and compelling biomimetic metal-organic framework catalyst for the degradation of nerve agent simulants. Angew. Chemie. **126**, 507–511 (2014). https://doi.org/10.1002/ange.201307520
98. McGuirk, C.M., Katz, M.J., Stern, C.L., Sarjeant, A.A., Hupp, J.T., Farha, O.K., Mirkin, C.A.: Turning on catalysis: incorporation of a hydrogen-bond-donating squaramide moiety into a Zr metal-organic framework. J. Am. Chem. Soc. **137**, 919–925 (2015). https://doi.org/10.1021/ja511403t
99. Sahoo, P.C., Jang, Y.N., Lee, S.W.: Enhanced biomimetic CO_2 sequestration and $CaCO_3$ crystallization using complex encapsulated metal organic framework. J. Cryst. Growth **373**, 96–101 (2013). https://doi.org/10.1016/j.jcrysgro.2012.11.043
100. Sasan, K., Lin, Q., Mao, C.Y., Feng, P.: Incorporation of iron hydrogenase active sites into a highly stable metal–organic framework for photocatalytic hydrogen generation. Chem. Commun. **50**, 10390–10393 (2014). https://doi.org/10.1039/c4cc03946g
101. Feng, Y., Chen, C., Liu, Z., Fei, B., Lin, P., Li, Q., Sun, S., Du, S.: Application of a Ni mercaptopyrimidine MOF as highly efficient catalyst for sunlight-driven hydrogen generation. J. Mater. Chem. A **3**, 7163–7169 (2015). https://doi.org/10.1039/c5ta00136f
102. Nath, I., Chakraborty, J., Verpoort, F.: Metal organic frameworks mimicking natural enzymes: a structural and functional analogy. Chem. Soc. Rev. **45**, 4127–4170 (2016). https://doi.org/10.1039/c6cs00047a
103. Li, Y., Xu, N., Zhu, W., Wang, L., Liu, B., Zhang, J., Xie, Z., Liu, W.: Nanoscale melittin@zeolitic imidazolate frameworks for enhanced anticancer activity and mechanism analysis. ACS Appl. Mater. Interfaces **10**, 22974–22984 (2018). https://doi.org/10.1021/acsami.8b06125
104. Wang, Q., Zhang, X., Huang, L., Zhang, Z., Dong, S.: GOx@ZIF-8(NiPd) nanoflower: an artificial enzyme system for tandem catalysis. Angew. Chemie. **129**, 16298–16301 (2017). https://doi.org/10.1002/ange.201710418
105. Wang, Y., Hou, C., Zhang, Y., He, F., Liu, M., Li, X.: Preparation of graphene nano-sheet bonded PDA/MOF microcapsules with immobilized glucose oxidase as a mimetic multi-enzyme system for electrochemical sensing of glucose. J. Mater. Chem. B **4**, 3695–3702 (2016). https://doi.org/10.1039/c6tb00276e
106. Dey, S., Mondal, B., Chatterjee, S., Rana, A., Amanullah, S., Dey, A.: Molecular electrocatalysts for the oxygen reduction reaction. Nat. Rev. Chem. **1** (2017). https://doi.org/10.1038/s41570-017-0098
107. Zhang, B., Sun, L.: Artificial photosynthesis: opportunities and challenges of molecular catalysts. Chem. Soc. Rev. **48**, 2216–2264 (2019). https://doi.org/10.1039/c8cs00897c
108. Zhao, Y.M., Yu, G.Q., Wang, F.F., Wei, P.J., Liu, J.G.: Bioinspired transition-metal complexes as electrocatalysts for the oxygen reduction reaction. Chem. - A Eur. J. **25**, 3726–3739 (2019). https://doi.org/10.1002/chem.201803764
109. Mihara, N., Yamada, Y., Takaya, H., Kitagawa, Y., Aoyama, S., Igawa, K., Tomooka, K., Tanaka, K.: Oxygen reduction to water by a cofacial dimer of iron(III)–porphyrin and iron(III)–phthalocyanine linked through a highly flexible fourfold rotaxane. Chem. - A Eur. J. **23**, 7508–7514 (2017). https://doi.org/10.1002/chem.201700082
110. Oliveira, R., Zouari, W., Herrero, C., Banse, F., Schöllhorn, B., Fave, C., Anxolabéhère-Mallart, E.: Characterization and subsequent reactivity of an Fe-peroxo porphyrin generated by electrochemical reductive activation of O_2. Inorg. Chem. **55**, 12204–12210 (2016). https://doi.org/10.1021/acs.inorgchem.6b01804
111. Liu, C., Lei, H., Zhang, Z., Chen, F., Cao, R.: Oxygen reduction catalyzed by a water-soluble binuclear copper(ii) complex from a neutral aqueous solution. Chem. Commun. **53**, 3189–3192 (2017). https://doi.org/10.1039/c6cc09206c
112. Kotani, H., Yagi, T., Ishizuka, T., Kojima, T.: Enhancement of 4-electron O_2 reduction by a Cu(II)-pyridylamine complex via protonation of a pendant pyridine in the second coordination sphere in water. Chem. Commun. **51**, 13385–13388 (2015). https://doi.org/10.1039/c5cc03012a

113. Zhang, W., Lai, W., Cao, R.: Energy-related small molecule activation reactions: oxygen reduction and hydrogen and oxygen evolution reactions catalyzed by porphyrin- and corrole-based systems. Chem. Rev. **117**, 3717–3797 (2017). https://doi.org/10.1021/acs.chemrev.6b00299
114. Sonkar, P.K., Prakash, K., Yadav, M., Ganesan, V., Sankar, M., Gupta, R., Yadav, D.K.: Co(II)-porphyrin-decorated carbon nanotubes as catalysts for oxygen reduction reactions: an approach for fuel cell improvement. J. Mater. Chem. A **5**, 6263–6276 (2017). https://doi.org/10.1039/c6ta10482g
115. Brezny, A.C., Nedzbala, H.S., Mayer, J.M.: Multiple selectivity-determining mechanisms of H_2O_2 formation in iron porphyrin-catalysed oxygen reduction. Chem. Commun. **57**, 1202–1205 (2021). https://doi.org/10.1039/d0cc06701f
116. Kostopoulos, N., Banse, F., Fave, C., Anxolabéhère-Mallart, E.: Modulating alkene reactivity from oxygenation to halogenationviaelectrochemical O_2 activation by Mn porphyrin. Chem. Commun. **57**, 1198–1201 (2021). https://doi.org/10.1039/d0cc07531k
117. Thiyagarajan, N., Janmanchi, D., Tsai, Y.-F., Wanna, W.H., Ramu, R., Chan, S.I., Zen, J.-M., Yu, S.S.-F.: A carbon electrode functionalized by a tricopper cluster complex: overcoming overpotential and production of hydrogen peroxide in the oxygen reduction reaction. Angew. Chemie. **130**, 3674–3678 (2018). https://doi.org/10.1002/ange.201712226
118. Lu, Y., Wang, X., Wang, M., Kong, L., Zhao, J.: 1,10-phenanthroline metal complex covalently bonding to poly- (pyrrole-3-carboxylic acid)-coated carbon: an efficient electrocatalyst for oxygen reduction. Electrochim. Acta **180**, 86–95 (2015). https://doi.org/10.1016/j.electacta.2015.08.104
119. Cunningham, D.W., Yang, J.Y.: Kinetic and mechanistic analysis of a synthetic reversible CO_2/HCO_2-electrocatalyst. Chem. Commun. **56**, 12965–12968 (2020). https://doi.org/10.1039/d0cc05556e
120. Kaeffer, N., Chavarot-Kerlidou, M., Artero, V.: Hydrogen evolution catalyzed by cobalt diimine-dioxime complexes. Acc. Chem. Res. **48**, 1286–1295 (2015). https://doi.org/10.1021/acs.accounts.5b00058
121. Xie, L., Tian, J., Ouyang, Y., Guo, X., Zhang, W., Apfel, U., Zhang, W., Cao, R.: Water-soluble polymers with appending porphyrins as bioinspired catalysts for the hydrogen evolution reaction. Angew. Chemie. **132**, 15978–15982 (2020). https://doi.org/10.1002/ange.202003836
122. Rosser, T.E., Gross, M.A., Lai, Y.H., Reisner, E.: Precious-metal free photoelectrochemical water splitting with immobilised molecular Ni and Fe redox catalysts. Chem. Sci. **7**, 4024–4035 (2016). https://doi.org/10.1039/c5sc04863j
123. Wu, H.L., Li, X.B., Tung, C.H., Wu, L.Z.: Bioinspired metal complexes for energy-related photocatalytic small molecule transformation. Chem. Commun. **56**, 15496–15512 (2020). https://doi.org/10.1039/d0cc05870j
124. Stieger, K.R., Ciornii, D., Kölsch, A., Hejazi, M., Lokstein, H., Feifel, S.C., Zouni, A., Lisdat, F.: Engineering of supramolecular photoactive protein architectures: the defined co-assembly of photosystem i and cytochrome: C using a nanoscaled DNA-matrix. Nanoscale **8**, 10695–10705 (2016). https://doi.org/10.1039/c6nr00097e
125. Guo, W., Hong, F., Liu, N., Huang, J., Wang, B., Duan, R., Lou, X., Xia, F.: Target-specific 3D DNA gatekeepers for biomimetic nanopores. Adv. Mater. **27**, 2090–2095 (2015). https://doi.org/10.1002/adma.201405078
126. Kocsis, I., Sorci, M., Vanselous, H., Murail, S., Sanders, S.E., Licsandru, E., Legrand, Y.M., Van der Lee, A., Baaden, M., Petersen, P.B., Belfort, G., Barboiu, M.: Oriented chiral water wires in artificial transmembrane channels. Sci. Adv. **4** (2018). https://doi.org/10.1126/sciadv.aao5603
127. Zhao, X.P., Liu, F.F., Hu, W.C., Younis, M.R., Wang, C., Xia, X.H.: Biomimetic nanochannel-ionchannel hybrid for ultrasensitive and label-free detection of microRNA in cells. Anal. Chem. **91**, 3582–3589 (2019). https://doi.org/10.1021/acs.analchem.8b05536
128. Burns, J.R., Seifert, A., Fertig, N., Howorka, S.: A biomimetic DNA-based channel for the ligand-controlled transport of charged molecular cargo across a biological membrane. Nat. Nanotechnol. **11**, 152–156 (2016). https://doi.org/10.1038/nnano.2015.279

129. Lv, C., Gu, X., Li, H., Zhao, Y., Yang, D., Yu, W., Han, D., Li, J., Tan, W.: Molecular transport through a biomimetic DNA channel on live cell membranes. ACS Nano **14**, 14616–14626 (2020). https://doi.org/10.1021/acsnano.0c03105
130. Hu, X.B., Liu, Y.L., Wang, W.J., Zhang, H.W., Qin, Y., Guo, S., Zhang, X.W., Fu, L., Huang, W.H.: Biomimetic graphene-based 3D scaffold for long-term cell culture and real-time electrochemical monitoring. Anal. Chem. **90**, 1136–1141 (2018). https://doi.org/10.1021/acs.analchem.7b03324
131. Liu, Y., Li, L., Zhu, J., Xu, J., Liu, S., Wang, Y., Zhang, C., Liu, T.: A biomimetic: Setaria viridis-inspired electrode with polyaniline nanowire arrays aligned on MoO_3@polypyrrole core-shell nanobelts. J. Mater. Chem. A **6**, 13428–13437 (2018). https://doi.org/10.1039/c8ta04218g
132. Wickham, A., Vagin, M., Khalaf, H., Bertazzo, S., Hodder, P., Dånmark, S., Bengtsson, T., Altimiras, J., Aili, D.: Electroactive biomimetic collagen-silver nanowire composite scaffolds. Nanoscale **8**, 14146–14155 (2016). https://doi.org/10.1039/c6nr02027e
133. Jiao, D., Lossada, F., Guo, J., Skarsetz, O., Hoenders, D., Liu, J., Walther, A.: Electrical switching of high-performance bioinspired nanocellulose nanocomposites. Nat. Commun. **12**, 1–10 (2021). https://doi.org/10.1038/s41467-021-21599-1
134. Li, B.M., Lu, J.: Cobalt in lithium-ion batteries. Sci. **367**, 970–980 (2020)
135. Shi, W., Shen, J., Shen, L., Hu, W., Xu, P., Baucom, J.A., Ma, S., Yang, S., Chen, X.M., Lu, Y.: Electrolyte membranes with biomimetic lithium-ion channels. Nano Lett. **20**, 5435–5442 (2020). https://doi.org/10.1021/acs.nanolett.0c01910
136. Chen, N., Dai, Y., Xing, Y., Wang, L., Guo, C., Chen, R., Guo, S., Wu, F.: Biomimetic ant-nest ionogel electrolyte boosts the performance of dendrite-free lithium batteries. Energy Environ. Sci. **10**, 1660–1667 (2017). https://doi.org/10.1039/c7ee00988g
137. Yabuuchi, N., Kajiyama, M., Iwatate, J., Nishikawa, H., Hitomi, S., Okuyama, R., Usui, R., Yamada, Y., Komaba, S.: P2-type $Na_x[Fe_{1/2}Mn_{1/2}]O_2$ made from earth-abundant elements for rechargeable Na batteries. Nat. Mater. **11**, 512–517 (2012). https://doi.org/10.1038/nmat3309
138. Liu, W., Zhi, H., Yu, X.: Recent progress in phosphorus based anode materials for lithium/sodium ion batteries. Energy Storage Mater. **16**, 290–322 (2019). https://doi.org/10.1016/j.ensm.2018.05.020
139. Ran, L., Luo, B., Gentle, I.R., Lin, T., Sun, Q., Li, M., Rana, M.M., Wang, L., Knibbe, R.: Biomimetic Sn_4P_3 anchored on carbon nanotubes as an anode for high-performance sodium-ion batteries. ACS Nano **14**, 8826–8837 (2020). https://doi.org/10.1021/acsnano.0c03432
140. Orita, A., Verde, M.G., Sakai, M., Meng, Y.S.: A biomimetic redox flow battery based on flavin mononucleotide. Nat. Commun. **7**, 1–8 (2016). https://doi.org/10.1038/ncomms13230
141. Soloveichik, G.L.: Flow batteries: current status and trends. Chem. Rev. **115**, 11533–11558 (2015). https://doi.org/10.1021/cr500720t
142. Dai, K., Wang, X., Yi, F., Jiang, C., Li, R., You, Z.: Triboelectric nanogenerators as self-powered acceleration sensor under high-g impact. Nano Energy **45**, 84–93 (2018). https://doi.org/10.1016/j.nanoen.2017.12.022
143. Liu, S., Liu, X., Zhou, G., Qin, F., Jing, M., Li, L., Song, W., Sun, Z.: A high-efficiency bioinspired photoelectric-electromechanical integrated nanogenerator. Nat. Commun. **11**, 1–9 (2020). https://doi.org/10.1038/s41467-020-19987-0
144. Wakerley, D., Lamaison, S., Ozanam, F., Menguy, N., Mercier, D., Marcus, P., Fontecave, M., Mougel, V.: Bio-inspired hydrophobicity promotes CO_2 reduction on a Cu surface. Nat. Mater. **18**, 1222–1227 (2019). https://doi.org/10.1038/s41563-019-0445-x

Progress in Bioelectrocatalysis

Graziela C. Sedenho

Abstract Oxidoreductases are highly specific and efficient biocatalysts that can be applied on electrochemical systems for green energy conversion, such as biological fuel cells, and in biosensors. However, understanding the biocatalytic and electron transfer mechanisms to reach high-performance devices is not trivial and has been requiring many efforts of researchers from bioelectrochemical field. In this context, the present chapter addresses in details the fundamental aspects and recent advances in the understanding of catalytic and electron transfer mechanisms of the main oxidoreductases applied electrochemical systems, such as alcohol dehydrogenase, hydrogenase, glucose oxidase, and multicopper oxidases.

1 Introduction

The core of the bioelectrocatalysis is the biocatalysts, and they can be distinguished by their nature in two main classes: microbial cells and redox enzymes, being the latter the focus of this chapter. The evolution process over billions of years contributed to the development of highly specific and efficient biocatalysts, so-called enzymes, that play crucial roles in the intricate metabolism of all living organisms on earth. In particular, oxidoreductases are a class of enzymes that play important role in energy conversion and metabolism in living organisms and represent almost 25% of all known proteins [1]. Those enzymes, through their cofactors or coenzymes, biologically catalyze redox reactions between two substrates, an electron donor and an electron acceptor. Several researchers have extensively demonstrated isolated oxidoreductases can be electrically connected to electrode surfaces, and the electrode is able to substitute one of the enzyme substrates.

In summary, the bioelectrocatalysis relies on the benefits of biocatalysis and electrocatalysis, as the biocatalysis is controlled by the electrode potential. The Fermi level of the electrode is controlled by the applied potential, and the electrode can

G. C. Sedenho (✉)
São Carlos Institute of Chemistry, University of São Paulo (USP), Avenida Trabalhador São-carlense, 400, São Paulo 13560-970, Brazil
e-mail: graziela.cs@usp.br

F. N. Crespilho (ed.), *Advances in Bioelectrochemistry Volume 2*,
https://doi.org/10.1007/978-3-030-95270-9_2

act as electron donor or receptor to/from the biocatalyst. In consequence, the electrochemical conversion of the substrate into product occurs at lower overpotentials than would take place without the biological entity. In addition, the biocatalysts are renewable, able to operate under mild conditions of pH, temperature, and pressure, and show wide substrate scope and higher activity and selectivity when compared to the traditional inorganic electrocatalysts [1, 2]. Therefore, these features of oxidoreductases enable practical applications of bioelectrocatalysis on several systems and devices, such as in biosensors [3], bioelectrosynthesis of diverse products and fuels [2, 4], and green energy conversion systems [5].

In this context, the present chapter addresses in details the types and fundamental aspects of the oxidoreductases mostly applied in energy conversion systems and biosensors and their catalytic and electron transfer mechanisms. Although microbial cells represent an important type of biocatalyst, the present chapter is focused on describing and reporting the recent progress in bioelectrocatalysis involving isolated redox enzymes.

2 Overall Aspects of the Redox Enzymes

Redox enzymes (or oxidoreductases) show a polypeptide backbone arranged in secondary and tertiary structures, and a prosthetic group, namely redox cofactor, which acts in the electron transfer process. The composition and conformation of the polypeptide backbone play crucial role in the biological recognition of the substrates and in the enzyme stability [6]. Typically, the redox cofactor can be metal complexes, such as heme centers, iron-sulfur clusters, and copper centers, or an organic molecule bound to a specific site of the enzyme, for example, quinones and flavins derivatives [7]. While some oxidoreductases show a cofactor very tightly or even covalently bound to their structure to catalyze the redox reaction, others require coenzymes as transient electron (or hydride) carriers. Examples of coenzymes of oxidoreductases are nicotinamide adenine dinucleotide (NAD) and nicotinamide adenine dinucleotide phosphate (NADP). Oxidoreductases can be divided according to the catalyzed redox reaction into dehydrogenases, oxygenases, oxidases, hydrogenases, and nitrogenases, and in a second level, they can be divided by the cofactor or coenzyme type [8].

Several oxidoreductases have been employed as bioelectrocatalysts in bioelectrosynthesis of value-added compounds and fuels, in biological fuel cells (BFCs), and in the development of biosensing devices. However, the present chapter is dedicated to describe the recent progress in the understanding of the bioelectrocatalytic and electron transfer mechanisms of the main dehydrogenases, hydrogenases, and oxidases used in bioelectrocatalytic systems.

Dehydrogenases represent the largest group of redox enzymes and are known to catalyze reversible hydride transfer reactions between their coenzymes or cofactors, such as NAD, flavin adenine dinucleotide (FAD), and quinone derivatives, and a diversity of organic alcohol, aldehydes, and sugars. Because of those features and their substrates, some dehydrogenases have been extensively used in anodes of BFCs and in biosensors mainly for food quality control, monitoring of industrial processes, and biomedical analyses [7]. Examples of these enzymes are alcohol dehydrogenase (ADH), glucose dehydrogenase (GDh), lactate dehydrogenase, fructose dehydrogenase, and malate dehydrogenase. Differently, hydrogenases are known to catalyze the reversible oxidation of molecular hydrogen into protons and electrons in their metal active sites, and because of that, they have extensively used as biocatalyst in bianodes of BFCs for hydrogen oxidation and in bioelectrosynthesis of H_2. The third class of redox enzymes addressed in the present chapter is the oxidases, which are able to transfer electron to oxygen and usually contain a flavin coenzyme or a metal cofactor as active site. One of the most studied and used oxidase in bioelectrochemical systems is glucose oxidase (GOD), a redox enzyme able to catalyze the oxidation of glucose into gluconolactone [9]. GOD is the most common oxidase used in bioanodes of BFCs and has been extensively used in glucose biosensors for diabetes diagnosis. Other very studied subclass of oxidases is the multicopper oxidases (MCOs), especially laccase and bilirubin oxidase (BOD). MCOs are able to promote the catalytic reduction of molecular oxygen into water and, because of that, they have been widely employed in biocathodes of BFCs [10]. The following subsections address aspects of bioelectrocatalytic and electron transfer mechanisms of those oxidoreductases.

3 Alcohol Dehydrogenase

ADH is a zinc-containing and NAD-dependent dehydrogenase that catalyzes the oxidation of a wide variety of primary and secondary alcohols to the corresponding aldehydes and ketones, respectively, with concomitant reduction of $NAD(P)^+$ to NAD(P)H. ADH is essential for the metabolism of endogenous and dietary ethanol in mammalians [11], whereas in fermentative microorganisms it inversely acts reducing acetaldehyde into ethanol. ADH from *Saccharomyces cerevisiae* mostly catalyzes the reversible oxidation of ethanol to acetaldehyde with the interconversion of NAD^+ to NADH (Eq. 1) and has been extensively used in the industrial production of ethanol and alcoholic beverages, and as anode catalyst of innumerous ethanol/O_2 enzymatic fuel cells [12].

$$CH_3CH_2OH + NAD^+ \rightleftarrows CH_3CHO + NADH + H^+ \quad (1)$$

Yeast ADH (EC 1.1.1.1) is an homotetramer with approximately 150 kDa (Fig. 1a). The asymmetric unit is formed by four different subunits arranged as similar dimers, AB and CD. Crystallographic studies have shown the unit cell contains two different tetramers consisting of AB:AB and CD:CD. The dimers have similar structures; however, the A and C subunits differ from the B and D subunits in coenzyme and catalytic domain structures. Each subunit in a dimer has a coenzyme binding domain, where the NAD binds to at the carboxyl-terminal group and two zinc atoms.

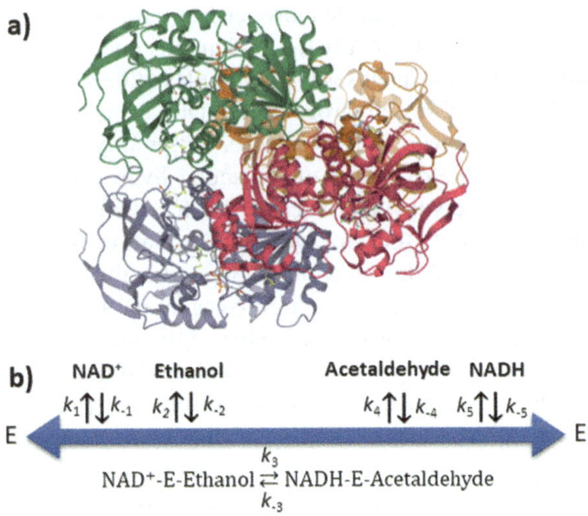

Fig. 1 **a** Yeast ADH structure from Protein Data Bank (PDB: 4W6Z). **b** Ordered sequential Bi-Bi catalytic mechanism of ethanol oxidation by AD. "E" represents the enzyme

One zinc atom is essential for catalytic activity, and the other has structural function. During catalysis, the alcohol binds in the cleft between the domains and coordinates to the tetrahedral zinc active site displacing water [13]. The tetrahedral catalytic zinc atoms are coordinated to two cysteine and one histidine residues, and then to the oxygen of the alcohol; therefore, the catalytic zinc atom promotes the precise positioning of the substrate to enable the hydride transfer from ethanol to NAD^+ [13].

It is well established that ADH shows a sequential bi-bi catalytic mechanism. NAD^+ binds first at the enzyme followed by ethanol binding, forming an ADH-ethanol-NAD^+ ternary complex. Then, the hydride is transferred producing ADH-acetaldehyde-NADH, and finally, the product dissociation of the product followed by the reduced coenzyme from the complex occurs [14]. More recently, studies with differential electrochemical mass spectrometry (DEMS) enabled further details about ethanol/acetaldehyde binding/dissociation kinetics and mechanism [15]. The concomitant detection of NADH and acetaldehyde showed that acetaldehyde dissociates from the enzyme structure approximately 10^7 times faster than NADH, confirming the ordered sequential Bi-Bi catalytic mechanism, and that the rate-limiting step is the dissociation of the NADH complex (Fig. 1b) [15].

4 Hydrogenase

Hydrogenases are crucial enzymes for many microorganisms that utilize molecular hydrogen as energy source [16]. In the bioelectrocatalysis field, they are known to efficiently catalyze the reversible oxidation of molecular hydrogen at small overpotentials in neutral media [17]. Since 2006 [18], due to their attractive features, purified hydrogenases have been widely exploited as biocatalyst in anodes of H_2/O_2 enzymatic fuel cells.

$$2H^+ + 2e^- \rightleftarrows H_2, E^{0'} = -0.413V \ (vs.SHE, pH7.0) \quad (2)$$

The several types of hydrogenases differ by the metal active site, which can be di-iron (FeFe), nickel–iron (NiFe), and iron (Fe), by the catalytic activity toward hydrogen oxidation or hydrogen evolution, and by O_2 tolerance. Between them, the [NiFe] hydrogenases, namely group 1, are the most studied type and show several advantages for application in anodes of H_2/O_2 enzymatic fuel cells, as they are more robust and show high active for hydrogen oxidation, with turnover frequencies reaching values >1,000 s^{-1} [17]. Several researchers have been dedicated to understand the structure and the catalytic and electron transfer mechanisms, as well as the activity inactivation by molecular oxygen in the last years, and in this context, crystallographic, spectroscopic, and electrochemical techniques have been provided useful information.

Group 1 hydrogenases show similar structures consisting of two subunits that interact very extensively forming a globular heterodimer [19]. The smaller subunit shows three Fe-S clusters, namely [4Fe-4S] proximal [3Fe-4S] medial, and [4Fe-4S] distal clusters, whereas the larger subunit houses the deeply buried NiFe active site (Fig. 2a). Although the Fe-S clusters do not participate in the catalytic mechanism, they play crucial role in the electron transfer mechanism, as they are aligned as a conductive chain that allows a fast electron transfer from the active site to the

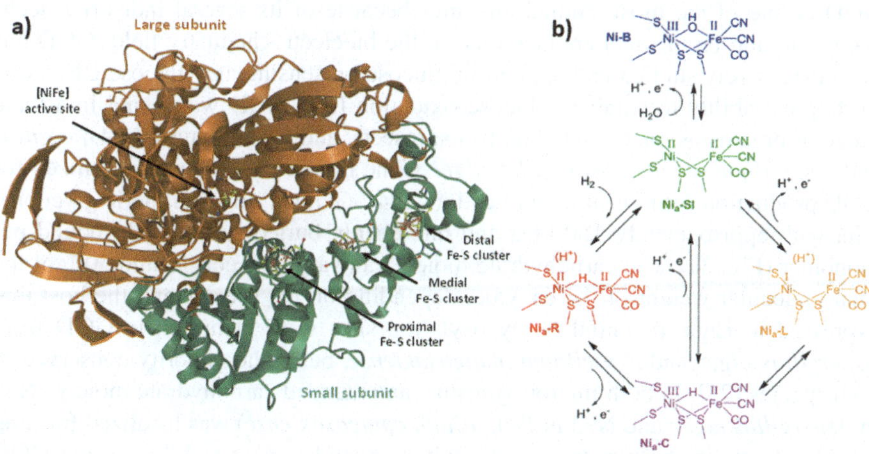

Fig. 2 **a** Structure of [NiFe] hydrogenases I from *E. coli* (PDB 3UQY) indicating the large and small units, the active site and Fe-S clusters. **b** Mechanism for H2 oxidation at [NiFe] hydrogenase. Reprinted from Ash, P.A., Hidalgo, R., Vincent, K.A.: Proton transfer in the catalytic cycle of [NiFe] hydrogenases: Insight from vibrational spectroscopy. ACS Catal. **7**, 2471–2485 (2017). https://pubs.acs.org/doi/abs/10.1021/acscatal.6b03182. Further permission related to this material should be directed to the ACS. Copyright 2017 American Chemical Society

enzyme surface, where they are transferred to an electron acceptor [17]. The catalytic bimetallic site has Ni and the Fe atoms bonded to the protein by four cysteine thiolates, two of which are bridging ligands, whereas the other two are terminal to the Ni, and Fe atom is further coordinated to one CO and two CN^- ligands. [19]

Despite variations between [NiFe] hydrogenases from different organisms and the localization in the microbial cells, they show a typical catalytic cycle, as shown in Fig. 2b. Considering the direction of molecular hydrogen oxidation, initially [NiFe] hydrogenase tends to be as a mixture of states at Ni(III)-Fe(II) (Ni–B state) and Ni(II)-Fe(II) (Ni_a-SI). Ni–B state requires a reductive activation, forming Ni_a-SI state which interacts with H_2, which is heterolytically cleaved [20, 21]. The H_2 cleavage leads to the formation of the Ni_a-R state with a bridging hydride ligand between Ni(II) and Fe(II) and a nearby proton. The removal of the proton generates the Ni_a-C state; then, the Ni_a-SI state is regenerated by a proton and electron transfer. There are spectroscopic evidences that an intermediate state involving Ni(I) (Ni_a-L state) may occur before reestablishing the initial Ni_a-SI state [22]. Despite several advances in unraveling the catalytic mechanism of hydrogenases through electrochemical control combined to spectroscopic monitoring of the catalytic site [20–23], some aspects remain unclear, such as the initial interaction between H_2 and Ni_a-SI, how heterolytic cleavage of H_2 proceeds, and how the proton leaves the active site.

5 Glucose Oxidase

GOD is one of the most studied enzymes because of its several industrial, technological, and biomedical applications. In the bioelectrochemistry field, GOD has been extensively studied and applied in glucose biosensing and glucose BFCs due to its great ability to catalyze glucose oxidation [24]. GOD was found in a wide range of organisms, including alga, fruits, insects, and fungi, mainly in *Aspergillus* and *Penicillium* species. *Aspergillus niger* is the most common species utilized for GOD production for industrial applications [24]. GOD is a homodimeric glycoprotein, with approximately 160 kDa and one deeply buried FAD tightly bound per monomer (Fig. 3a). The carbohydrate moiety can vary between 11 and 30% of the total molecular weight of the enzyme, depending on the species and the host that expressed it. Eight potential N-glycosylation sites were identified in GOD from *Aspergillus niger* and *Penicillium amagasakiense*, being the majority conserved in both species [25]. In both microorganisms, an extended carbohydrate moiety (N89 in *Aspergillus niger* and N93 in *Penicillium amagasakiense*) was localized forming a bridge between the two monomers, which provides extra stability of the GOD dimer. The FAD-binding site is quite conserved in GOD from *Aspergillus niger* and *Penicillium amagasakiense;* all residues forming hydrogen bonds to the cofactor are conserved, except His78 and Thr110 in the protein from *Aspergillus niger*, which are replaced by Gln78 and Ser110 in GOD from *Penicillium amagasakiense*, respectively [25].

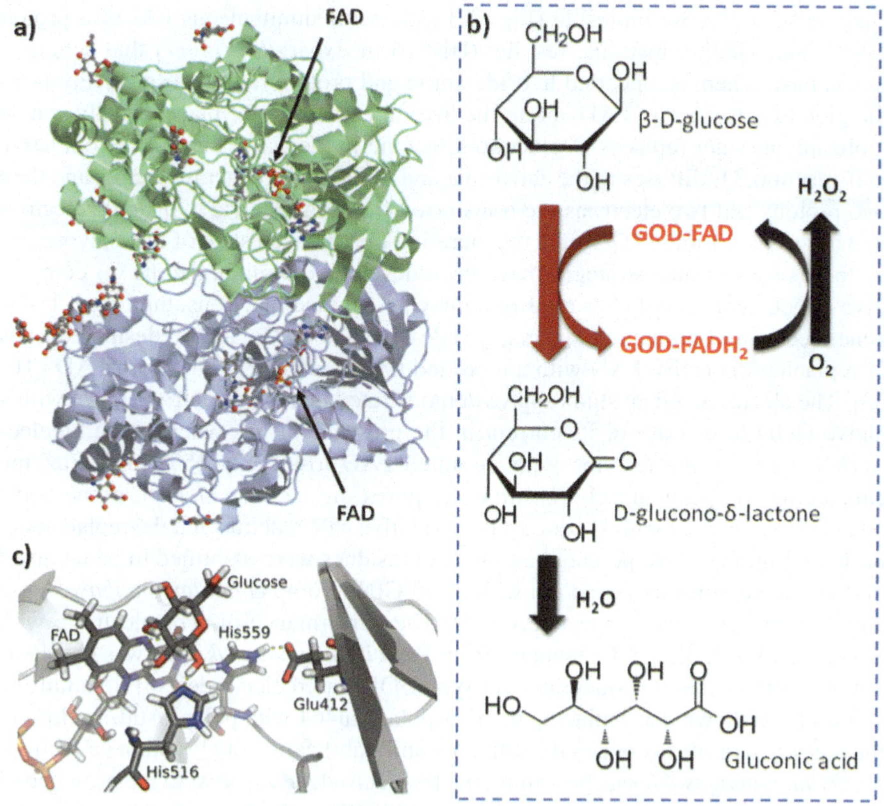

Fig. 3 **a** Structure of GOD from *Penicillium amagasakiense* (PDB 1GPE). **b** Scheme of the glucose oxidation by GOD, with regeneration of FAD by electron transfer to molecular oxygen. **c** The active site of GOD from *Aspergillus niger* with bonded glucose molecule. Carbon, nitrogen, oxygen, hydrogen, and phosphorous atoms are represented in gray, blue, red, white, and orange, respectively. Reprinted from Petrović, D., Frank, D., Kamerlin, S.C.L., Hoffmann, K., Strodel, B.: Shuffling active site substate populations affects catalytic activity: The case of glucose oxidase. ACS Catal. **7**, 6188–6197 (2017). https://pubs.acs.org/doi/10.1021/acscatal.7b01575. Further permission related to this material should be directed to the ACS. Copyright 2017 American Chemical Society

The function of this flavoprotein is to catalyze the oxidation of glucose. This process is divided into two sequential half-reactions: firstly, the β-D-glucose is oxidized to D-glucono-δ-lactone, through the transfer of two protons and two electrons from the substrate to FAD, in a reductive half-reaction. The D-glucono-δ-lactone formed is then non-enzymatically hydrolyzed to gluconic acid (Fig. 3b). Next, in the oxidative half-reaction, O_2, the natural co-substrate of the enzyme, is reduced to H_2O_2 for regeneration of the oxidized cofactor, as represented in Fig. 3b [26]. Since the 1970, it is established that glucose oxidation reaction by GOD follows a Ping-Pong Bi-Bi mechanism [27–29]. The first step of the catalytic mechanism involves the replacing of a water molecule from the oxidized active site of the enzyme by a glucose molecule (Fig. 3c) [30], with simultaneous loss of a proton from a surrounding histidine residue (His516 in *Aspergillus niger*) that acts as a Lewis base. Then, a concerted hydride anion and proton transfer respectively from the glucose substrate to FAD and to the histidine residue take place. Finally, a new molecule of water replaces D-glucono-δ-lactone in the active site. In the oxidative half-reaction, O_2 diffuses to the active site and replaces the water molecule, and then two protons and two electrons are transferred from $FADH_2$ via two single electron transfer steps, forming H_2O_2 and regenerating the oxidized state of the enzyme.

In this sense, some strategies have been developed to minimize the O_2 competitive effect, such as cofactor redesign, enzyme engineering, consumption of H_2O_2 generated, use of protection layers and redox mediators. Cofactor redesign consists of replacing the native FAD with a modified flavin, such as 7,8-dichloro-FAD [31–33]. The electrochemical studies have demonstrated that a bioelectrode made with a native GOD loses 60% of its current in the presence of O_2, whereas the bioelectrode employing the enzyme with modified FAD loses only 15% [33]. Enzyme engineering for reducing O_2-sensitivity requires the identification of amino acids residues that interact with O_2 during the oxidative half-reaction and the replacement of them. Eight and five potential amino acid residues were identified to be involved during the oxidative half-reaction with O_2 in GOD from *Penicilium amagasakiense* and *Aspergillus niger*, respectively [34]. The performance of a bioelectrode with Ser114Ala/Phe355Leu GOD mutant from *Penicilium amagasakiense* was considerably less affected by O_2 than the wild-type GOD-based electrode [35]. In addition, studies have shown that replacing the non-polar Val464 with polar histidine close to the redox active site decreases the diffusion and stabilization of O_2 within GOD from *Penicilium amagasakiense*, becoming the bioelectrode insensitive to the presence of oxygen and considerably more stable due to the lower amount of H_2O_2 generated [36]. Other strategies to circumvent the effect of H_2O_2 on GOD bioelectrochemical systems are related to enzyme immobilization. Co-immobilization of catalase [30, 37–39], horseradish peroxidase (HRP) [40], and polydopamine [41] for decomposition of the H_2O_2, as well as the use a protection layer consisted of lactate oxidase and catalase to consume O_2 and avoid its contact with GOD catalytic layer [42] are able to provide O_2 tolerant GOD bioelectrodes. Alternatively, H_2O_2 generated by GOD on the bioanode has been used to supply the HRP biocatalyst on the cathode of a glucose BFC [43].

Another important aspect that represents a recurrent difficulty in the development of GOD-based bioelectrochemical systems is the efficient electrical communication between the active site of the enzyme and the electrode surface, to achieve high electron transfer rates and consequently high catalytic currents. As previously mentioned, the redox sites (FAD molecules) of GOD are deeply buried into the polypeptide shell of the enzyme, which hampers DET to the electrode surface by electron tunneling. Some strategies have emerged to decrease the distance between FAD, such as the use of redox polymers and nanomaterials and orientated enzyme immobilization, and enzyme modification. Since the 1980 [46, 47], redox mediators have been used to decrease the distance that electrons need to travel from the redox active site of GOD to the electrode surface, increasing the rate of tunneling and consequently the bioelectrode performance. Examples of redox mediators are: osmium-based polymers [46, 48], ferrocene derivatives [47, 49, 50], ferritin [51, 52], quinone derivatives [45], and redox dye-based polymers [44]. Also, nanomaterials such as gold [53–55] and platinum nanoparticles [56], carbon nanotubes [57, 58], and graphene oxide [5] have been applied with this purpose. Alternatively, GOD deglycosylation [59, 60] and oligomerization [61] have been used to enhance the electron transfer reaction between FAD and the electrode surface. The high content of carbohydrate moiety on GOD acts insulating for electrical contact of the redox active site; therefore, the enzyme deglycosylation can short the redox site/electrode distance compared to the native enzyme. GOD oligomerization can improve the interaction of the enzyme with the electrode surface, considerably enhancing the electron transfer rate when compared to the native protein.

6 Multicopper Oxidases

MCO is a family of oxidases ubiquitous in the nature and can be divided into three categories according to the specificity toward the substrates. The first category includes laccases, which oxidize a wide range of organic substrates; the second one is the metalloxidases, such as ceruloplasmin and cuprous oxidase, that oxidize metal substrates; and the third category is represented by enzymes that oxidize specific organic substrates, for example, BOD and ascorbate oxidase [10]. The overall feature of the MCOs is the presence of at least four copper atoms in their structure, which act in the substrate oxidation while transfer electrons to the molecular oxygen, their natural electron acceptor.

Laccase was the first enzyme used for reducing O_2, in 1979 [62]. Although BOD was discovered and isolated in 1981 from *Myrothecium verrucaria* MT-1 fungal [63], its potential for reducing O_2 was demonstrated only in 2001, starting the development of BOD-based biocathodes for BFCs [64]. In 2011, the crystallographic structure of BOD from *Myrothecium verrucaria* was completely elucidated [65]. It is known BOD (EC 1.3.3.5) is a glycosylated monomeric enzyme with molecular weight ranging from 52 to 68 kDa, 534 amino acid residues, and four copper ions (Fig. 4a) [65]. Although laccases have been extensively studied and employed in biocathodes for reducing O_2, BODs become much more attractive for that because they show higher catalytic activity in neutral media and are less inhibited by halides than laccases [66, 67].

The catalytic site of MCOs is formed by the arrangement of the four copper ions, namely Type 1 (T1), Type 2 (T2), and two copper ions Type 3 (T3). The Cu T2 and T3 (two ions) form a trinuclear cluster (TNC), which is located 1.2–1.3 nm far from T1 [65]. The T1 and T3 Cu atoms are linked by a sequence of three amino acid residues (His-Cys-His) [68], that are conserved in MCOs, and provide a fast intramolecular electron-transfer pathway. The catalytic mechanism of the MCO has been widely studied by spectroscopic, crystallographic, and density-functional theory. It is known that the oxidation of organic substrates and reduction of O_2 follow

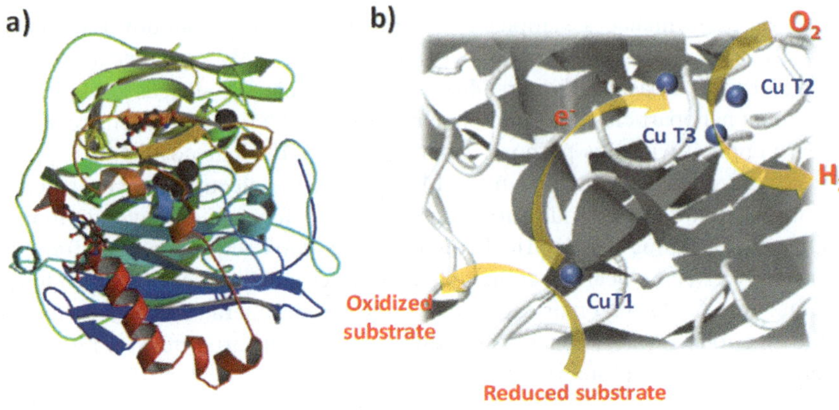

Fig. 4 **a** Structure of BOD from *Myrothecium verrucaria* MT-1 (PDB: 2XLL). **b** Catalytic site of MCOs with the arrows indicating the flow of electrons, substrate, and O_2, and the blue spheres indicating the copper ions

a ping-pong mechanism. Summarily, T1 Cu is located near the protein surface and is the primary electron acceptor from the natural substrates. Then, the electrons are transferred, one at a time, from T1 to the T2/T3 Cu cluster, where O_2 is reduced to water (Fig. 4b) [69–71].

Detailed chemical, spectroscopic, and computational studies of ORR catalyzed by MCOs greatly contributed to elucidate the catalytic mechanism of those enzymes [68–71]. It is known that the mechanism involves the formation of two highly reactive intermediates and occurs in two 2-electron steps with the second being fast and ensuring a four-electron process, in fact. The catalytic mechanism can be described as follows: (1) the fully reduced native enzyme (all Cu ions in the Cu(I) state) reacts with O_2 and 2 electrons are transferred, forming a peroxy-intermediate (PI), where O-O bond is conserved. In the PI, O_2 forms a bridge between T2 Cu and T3 Cu, being that T2 Cu and one T3 Cu are in Cu(II) state, no protons are involved in the formation of PI. Then, (2) a 2-electron transfer occurs and PI is converted into a native intermediate (NI). In the NI, all copper ions are in the Cu(II) state. The O-O bond is cleaved and the oxygen atoms originally from the O_2 molecule form an oxo and a hydroxo bridge between T2 Cu and T3 Cu, and between the two T3 Cu ions, respectively. (3) Lastly, in the presence of the electron donor substrate, the fully reduced state of the enzyme is reestablished by the transfer of four electrons and four protons, and releasing two H_2O molecules [70]. Therefore, NI is actually the catalytically relevant fully oxidized form of the MCOs. Studies with mutant BODs indicated that an aspartate residue in the TNC nearby conserved in all MCOs plays a crucial role in the proton supply for the RRO to water. In the low potential MCOs, the T1 reduction by substrate is the rate-limiting, whereas the rate-limiting step in turnover of high potential MCOs is the first intramolecular electron transfer to NI [71].

The redox potentials of the copper sites of MCOs were determined by redox titration and protein film voltammetry in anaerobic conditions. For BOD, the reduction potentials of T1 and T2/T3 were determined to be, respectively, between 0.66 V and 0.67 V vs. SHE (standard hydrogen electrode), and between 0.38 V and 0.39 V vs. SHE, at pH 7.0 [72–74].

In bioelectrochemical systems, the natural electron donor substrates of the MCOs are replaced by the electrode (DET mechanism) or redox mediators (MET), which further transfer electrons to the electrode surface. Several bioelectrochemical studies have contributed to the understanding of the electron transfer and biocatalytic mechanisms of MCOs. Electrochemical studies have demonstrated the current-pH dependency of BOD reflects the pK_a of the aspartate residue close to the TNC, which is involved in proton transfer during the reduction of O_2 to water. In addition, at pH close to the aspartate pK_a the BOD bioelectrocatalytic rate is controlled by the electrode potential that drives the T1 Cu reduction [66]. Therefore, the decrease of O_2 catalytic reduction currents with the electrolyte pH has been explained by the difficulty of supplying protons by such aspartate residue in question [66].

Electrochemical studies with laccase and BOD adsorbed on electrode surfaces have shown the enzymes can adopt different orientations on carbon-based electrodes and bare gold electrode, which can drastically affect the electron transfer efficiency and the bioelectrocatalysis of O_2 reduction. Carbon-based electrodes lead to the most enzyme molecules oriented with the Cu T1 site proximal to the electrode, favoring the DET-based bioelectrocatalysis, whereas gold electrodes induce the orientation with Cu T2/T3 cluster closer to the electrode, harming the DET bioelectroreduction of O_2 to water. It seems that DET from the electrode to the T2/T3 is possible but leads either to no catalysis or to reduction of O_2 via 2 electrons producing H_2O_2 [75, 76].

Recently, XAS studies under bioelectrochemical control (Fig. 5a) have significantly contributed to the understanding of some aspects of the BOD electron transfer mechanism [77]. Through operando XAS, it was demonstrated that an overpotential of 150 mV is required to reduce the four Cu(II) ions of the BOD in the presence of O_2 when compared to in anaerobic conditions. Although the Cu(II) ions are reduced at a potential close to + 0.55 V (vs. Ag/AgCl/KCl_{sat}) during the electrocatalytic ORR, they do not remain in the reduced oxidation state of +1 due to the presence of oxygen in the reactive site, and the electrons are quickly transferred in a second reaction step to the O_2 coordinated to the Cu TNC, hence returning to the +2-oxidation state, as schematically represented in Fig. 5b. Also, operando XAS combined with quantum mechanics/molecular mechanics suggests the metallic cofactor electronic structure, as well as the spatial disposition of its surrounding ligands, strongly influences the electron transfer that promotes the biocatalysis [78].

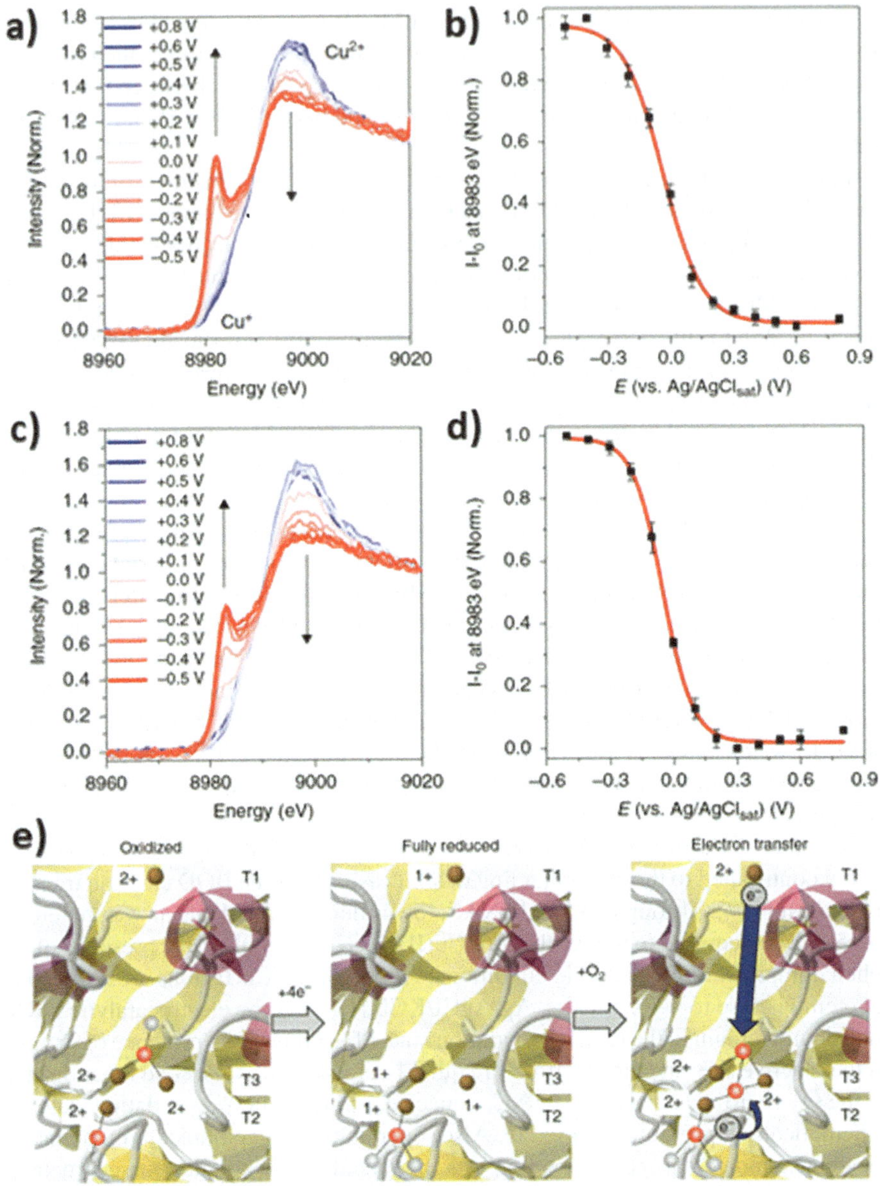

Fig. 5 Cu K-edge XAS spectra at different applied potentials in the absence (**a**) and presence (**c**) of O_2. Signals at 8983 eV for reductive titrations in the absence (**b**) and presence (**d**) of O_2. Red lines in panels **b** and **d** represent the fits using the Nernst equation. (**e**) Simplified and schematic mechanism of internal electron transfer in BOD in the presence of O_2. Reprinted from Macedo, L.J.A., Hassan, A., Sedenho, G.C., Crespilho, F.N.: Assessing electron transfer reactions and catalysis in multicopper oxidases with operando X-ray absorption spectroscopy. Nat. Commun. **11**, 316 (2020). Copyright 2020, Springer Nature

7 Final Considerations

In recent decades, research in enzymatic bioelectrocatalysis has made great progress, mainly regarding the understanding of biocatalytic and electron transfer mechanisms. This only was possible to the electrochemical, spectroscopic, and computational studies, which are often strategically combined to give detailed and multi-faceted information on the biomolecules. The understanding of the fundamental aspects of bioelectrocatalysis and electron transfer is not only important to understand the redox reactions in living organisms, but also has direct impact on the obtention of high-performance bioelectrochemical systems, such as biosensors, energy conversion systems and bioelectrosynthesis, and in the development of efficient biomimetic catalysts.

References

1. Chen, H., Simoska, O., Lim, K., Grattieri, M., Yuan, M., Dong, F., Lee, Y.S., Beaver, K., Weliwatte, S., Gaffney, E.M., Minteer, S.D.: Fundamentals, applications, and future directions of bioelectrocatalysis. Chem. Rev. **120**, 12903–12993 (2020). https://doi.org/10.1021/acs.chemrev.0c00472
2. Chen, H., Dong, F., Minteer, S.D.: The progress and outlook of bioelectrocatalysis for the production of chemicals, fuels and materials. Nat. Catal. **3**, 225–244 (2020). https://doi.org/10.1038/s41929-019-0408-2
3. Teymourian, H., Barfidokht, A., Wang, J.: Electrochemical glucose sensors in diabetes management: an updated review (2010–2020). Chem. Soc. Rev. **49**, 7671–7709 (2020). https://doi.org/10.1039/d0cs00304b
4. Adachi, T., Kitazumi, Y., Shirai, O., Kano, K.: Recent progress in applications of enzymatic bioelectrocatalysis. Catalysts **10**, 1–21 (2020). https://doi.org/10.3390/catal10121413
5. Martins, M.V.A., Pereira, A.R., Luz, R.A.S., Iost, R.M., Crespilho, F.N.: Evidence of short-range electron transfer of a redox enzyme on graphene oxide electrodes. Phys. Chem. Chem. Phys. **16**, 17426–17436 (2014). https://doi.org/10.1039/C4CP00452C
6. Nelson, D.L., Cox, M.M.: Lehninger Principles of Biochemistry, 5th edn. W. H. Freeman & Company, New York (2008)
7. Pereira, A.R., Sedenho, G.C., de Souza, J.C.P., Crespilho, F.N.: Advances in enzyme bioelectrochemistry. An. Acad. Bras. Cienc. **90**, 825–857 (2018). https://doi.org/10.1590/0001-3765201820170514
8. Blank, L.M., Ebert, B.E., Buehler, K., Bühler, B.: Redox biocatalysis and metabolism: molecular mechanisms and metabolic network analysis. Antioxid. Redox Signal **13**, 349–394 (2010). https://doi.org/10.1089/ars.2009.2931
9. Cosnier, S., Gross, A.J., Giroud, F., Holzinger, M.: Beyond the hype surrounding biofuel cells: what's the future of enzymatic fuel cells? Curr. Opin. Electrochem. **12**, 148–155 (2018). https://doi.org/10.1016/j.coelec.2018.06.006
10. Mano, N., de Poulpiquet, A.: O_2 reduction in enzymatic biofuel cells. Chem. Rev. **118**, 2392–2468 (2017). https://doi.org/10.1021/acs.chemrev.7b00220
11. Edenberg, H.J., Bosron, W.F.: Alcohol dehydrogenases. In: McQueen, C.A. (ed.) Comprehensive Toxicology, 2nd edn., pp. 111–130. Elsevier Ltd (2010)
12. Luz, R.A.S., Pereira, A.R., de Souza, J.C.P., Sales, F.C.P.F., Crespilho, F.N.: Enzyme biofuel cells: thermodynamics, kinetics and challenges in applicability. ChemElectroChem **1**, 1751–1777 (2014). https://doi.org/10.1002/celc.201402141

13. Raj, S.B., Ramaswamy, S., Plapp, B.V.: Yeast alcohol dehydrogenase structure and catalysis. Biochem. **53**, 5791–5803 (2014). https://doi.org/10.1021/bi5006442
14. Dickinson, F.M., Monger, G.P.: A study of the kinetics and mechanism of yeast alcohol dehydrogenase with a variety of substrates. Biochem. J. **131**, 261–270 (1973). https://doi.org/10.1042/bj1310261
15. de Souza, J.C.P., Silva, W.O., Lima, F.H.B., Crespilho, F.N.: Enzyme activity evaluation by differential electrochemical mass spectrometry. Chem. Commun. **53**, 8400–8402 (2017). https://doi.org/10.1039/C7CC03963H
16. Vignais, P.M., Billoud, B.: Occurrence, classification, and biological function of hydrogenases: an overview. Chem. Rev. **107**, 4206–4272 (2007). https://doi.org/10.1021/cr050196r
17. Lojou, E.: Hydrogenases as catalysts for fuel cells: strategies for efficient immobilization at electrode interfaces. Electrochim. Acta **56**, 10385–10397 (2011). https://doi.org/10.1016/j.electacta.2011.03.002
18. Vincent, K.A., Cracknell, J.A., Clark, J.R., Ludwig, M., Lenz, O., Friedrich, B., Armstrong, F.A.: Electricity from low-level H_2 in still air—an ultimate test for an oxygen tolerant hydrogenase. Chem. Commun., 5033–5035 (2006). https://doi.org/10.1039/b614272a
19. Volbeda, A., Amara, P., Darnault, C., Mouesca, J., Parkin, A., Roessler, M.M.: X-ray crystallographic and computational studies of the O_2 -tolerant [NiFe]-hydrogenase 1 from Escherichia coli. Proc. Natl. Acad. Sci. U. S. A. **109**, 5305–5310 (2012). https://doi.org/10.1073/pnas.1119806109
20. Ash, P.A., Kendall-price, S.E.T., Vincent, K.A.: Unifying activity, structure, and spectroscopy of [NiFe] hydrogenases: combining techniques to clarify mechanistic understanding. Acc. Chem. Res. **52**, 3120–3131 (2019). https://doi.org/10.1021/acs.accounts.9b00293
21. Ash, P.A., Hidalgo, R., Vincent, K.A.: Proton transfer in the catalytic cycle of [NiFe] hydrogenases: insight from vibrational spectroscopy. ACS Catal. **7**, 2471–2485 (2017). https://doi.org/10.1021/acscatal.6b03182
22. Hidalgo, R., Ash, P.A., Healy, A.J., Vincent, K.A.: Infrared spectroscopy during electrocatalytic turnover reveals the Ni-L active site state during H_2 oxidation by a NiFe hydrogenase. Angew. Chem. – Int. Ed. **54**, 7110–7113 (2015). https://doi.org/10.1002/anie.201502338
23. Evans, R.M., Ash, P.A., Beaton, S.E., Brooke, E.J., Vincent, K.A., Carr, S.B., Armstrong, F.A.: Mechanistic exploitation of a self-repairing, blocked proton transfer pathway in an O_2-tolerant [NiFe]-hydrogenase. J. Am. Chem. Soc. **140**, 10208–10220 (2018). https://doi.org/10.1021/jacs.8b04798
24. Mano, N.: Engineering glucose oxidase for bioelectrochemical applications. Bioelectrochemistry **128**, 218–240 (2019). https://doi.org/10.1016/j.bioelechem.2019.04.015
25. Wohlfahrt, G., Witt, S., Hendle, J., Schomburg, D., Kalisz, H.M., Hecht, H.-J.J.: 1.8 and 1.9 Å resolution structures of the Penicillium amagasakiense and Aspergillus niger glucose oxidases as a basis for modelling substrate complexes. Acta Crystallogr. Sect. D. Biol. Crystallogr. **55**, 969–977 (1999). https://doi.org/10.1107/S0907444999003431
26. Bankar, S.B., Bule, M.V., Singhal, R.S., Ananthanarayan, L.: Glucose oxidase—an overview. Biotechnol. Adv. **27**, 489–501 (2009). https://doi.org/10.1016/j.biotechadv.2009.04.003
27. Gibson, Q.H., Swoboda, B.E., Massey, V.: Kinetics and mechanism of action of glucose oxidase. J. Biol. Chem. **239**, 3927–3934 (1964). https://doi.org/10.1016/s0021-9258(18)91224-x
28. Weibel, M.K., Bright, H.J.: The glucose oxidase mechanism. Interpretation of the pH dependence. J. Biol. Chem. **246**, 2734–2744 (1971). https://doi.org/10.1016/S0021-9258(18)62246-X
29. Bright, H.J., Appleby, M.: The pH dependence of the individual steps in the glucose oxidase reaction. J. Biol. Chem. **244**, 3625–3634 (1969). https://doi.org/10.1016/s0021-9258(18)83415-9
30. Petrović, D., Frank, D., Kamerlin, S.C.L., Hoffmann, K., Strodel, B.: Shuffling active site substate populations affects catalytic activity: the case of glucose oxidase. ACS Catal. **7**, 6188–6197 (2017). https://doi.org/10.1021/acscatal.7b01575
31. Milton, R.D., Giroud, F., Thumser, A.E., Minteer, S.D., Slade, R.C.T.: Hydrogen peroxide produced by glucose oxidase affects the performance of laccase cathodes in glucose/oxygen

fuel cells: FAD-dependent glucose dehydrogenase as a replacement. Phys. Chem. Chem. Phys. **15**, 19371–19379 (2013). https://doi.org/10.1039/c3cp53351d
32. Christwardana, M., Chung, Y., Kwon, Y.: Co-immobilization of glucose oxidase and catalase for enhancing the performance of a membraneless glucose biofuel cell operated under physiological conditions. Nanoscale **9**, 1993–2002 (2017). https://doi.org/10.1039/c6nr09103b
33. Roth, J.P., Wincek, R., Nodet, G., Edmondson, D.E., McIntire, W.S., Klinman, J.P.: Oxygen isotope effects on electron transfer to O_2 probed using chemically modified flavins bound to glucose oxidase. J. Am. Chem. Soc. **126**, 15120–15131 (2004). https://doi.org/10.1021/ja047050e
34. Courjean, O., Hochedez, A., Neri, W., Louërat, F., Tremey, E., Gounel, S., Tsujimura, S., Mano, N.: A two-step synthesis of 7,8-dichloro-riboflavin with high yield. RSC Adv. **2**, 2700–2701 (2012). https://doi.org/10.1039/c2ra01211a
35. Tremey, E., Suraniti, E., Courjean, O., Gounel, S., Stines-Chaumeil, C., Louerat, F., Mano, N.: Switching an O_2 sensitive glucose oxidase bioelectrode into an almost insensitive one by cofactor redesign. Chem. Commun. **50**, 5912–5914 (2014). https://doi.org/10.1039/c4cc01670j
36. Horaguchi, Y., Saito, S., Kojima, K., Tsugawa, W., Ferri, S., Sode, K.: Construction of mutant glucose oxidases with increased dye-mediated dehydrogenase activity. Int. J. Mol. Sci. **13**, 14149–14157 (2012). https://doi.org/10.3390/ijms131114149
37. Horaguchi, Y., Saito, S., Kojima, K., Tsugawa, W., Ferri, S., Sode, K.: Engineering glucose oxidase to minimize the influence of oxygen on sensor response. Electrochim. Acta **126**, 158–161 (2014). https://doi.org/10.1016/j.electacta.2013.09.018
38. Tremey, E., Stines-Chaumeil, C., Gounel, S., Mano, N.: Designing an O_2-insensitive glucose oxidase for improved electrochemical applications. ChemElectroChem **4**, 2520–2526 (2017)
39. Cinquin, P., Gondran, C., Giroud, F., Mazabrard, S., Pellissier, A., Boucher, F.F., Alcaraz, J.P., Gorgy, K., Lenouvel, F., Mathé, S., Porcu, P., Cosnier, S.: A glucose BioFuel cell implanted in Rats. PLoS ONE **5**, 1–7 (2010). https://doi.org/10.1371/journal.pone.0010476
40. Zebda, A., Cosnier, S., Alcaraz, J.-P., Holzinger, M., Le, G.A., Gondran, C., Boucher, F., Giroud, F., Gorgy, K., Lamraoui, H., Cinquin, P.: Single glucose biofuel cells implanted in rats power electronic devices. Sci. Rep. **3**, 1516 (2013). https://doi.org/10.1038/srep01516
41. El, I.-R., Alcaraz, J., Boucher, F., Boutaud, B., Dalmolin, R., Boutonnat, J., Cinquin, P., Zebda, A., Martin, D.K.: Remote wireless control of an enzymatic biofuel cell implanted in a rabbit for 2 months. Electrochim. Acta **269**, 360–366 (2018). https://doi.org/10.1016/j.electacta.2018.02.156
42. Yimamumaimaiti, T., Lu, X., Zhang, J.R., Wang, L., Zhu, J.J.: Efficient blood-toleration enzymatic biofuel cell via in situ protection of an enzyme catalyst. ACS Appl. Mater. Interfaces **12**, 41429–41436 (2020). https://doi.org/10.1021/acsami.0c11186
43. Chen, H., Ru, X., Wang, H., Liu, P., Li, G., Cao, Y., Bai, Z., Yang, L.: Construction of a cascade catalyst of nanocoupled living red blood cells for implantable biofuel cell. ACS Appl. Mater. Interfaces **13**, 28010–28016 (2021). https://doi.org/10.1021/acsami.1c01479
44. Lopez, F., Zerria, S., Ruff, A., Schuhmann, W.: An O_2 tolerant polymer/glucose oxidase based bioanode as basis for a self-powered glucose sensor. Electroanal. **30**, 1311–1318 (2018). https://doi.org/10.1002/elan.201700785
45. Abreu, C., Nedellec, Y., Ondel, O., Buret, F., Cosnier, S., Le Goff, A., Holzinger, M.: Glucose oxidase bioanodes for glucose conversion and H_2O_2 production for horseradish peroxidase biocathodes in a flow through glucose biofuel cell design. J. Power Sources **392**, 176–180 (2018). https://doi.org/10.1016/j.jpowsour.2018.04.104
46. Degani, Y., Heller, A.: Electrical communication between redox centers of glucose oxidase and electrodes via electrostatically and covalently bound redox polymers. J. Am. Chem. Soc. **111**, 2357–2358 (1989)
47. Cass, A.E.G., Davis, G., Francis, G.D., Hill, H.A.O., Aston, W.J., Higgins, I.J., Plotkin, E.V., Scott, L.D.L., Turner, A.P.F.: Ferrocene-mediated enzyme electrode for amperometric determination of glucose. Anal. Chem. **56**, 667–671 (1984)
48. Yin, S., Liu, X., Kobayashi, Y., Nishina, Y., Nakagawa, R., Yanai, R., Kimura, K., Miyake, T.: A needle-type biofuel cell using enzyme/mediator/carbon nanotube composite fibers for

wearable electronics. Biosens. Bioelectron. **165**, 112287 (2020). https://doi.org/10.1016/j.bios.2020.112287
49. Schuhmann, W., Ohara, T.J., Schmidt, H.L., Heller, A.: Electron transfer between glucose oxidase and electrodes via redox mediators bound with flexible chains to the enzyme surface. J. Am. Chem. Soc. **113**, 1394–1397 (1991)
50. Lee, D., Jeong, S.H., Yun, S., Kim, S., Sung, J., Seo, J., Son, S., Kim, J.T., Susanti, L., Jeong, Y., Park, S., Seo, K., Kim, S.J., Chung, T.D. Totally implantable enzymatic biofuel cell and brain stimulator operating in bird through wireless communication. Biosens. Bioelectron. **171**, 112746 (2021). https://doi.org/10.1016/j.bios.2020.112746
51. ul Haque, S., Inamuddin, Nasar, A., Asiri, A.M.: Fabrication and characterization of electrochemically prepared bioanode (polyaniline/ferritin/glucose oxidase) for biofuel cell application. Chem. Phys. Lett. **692**, 277–284 (2018). https://doi.org/10.1016/j.cplett.2017.12.035
52. ul Haque, S., Nasar, A., Inamuddin, Rahman, M.M.: Applications of chitosan (CHI)-reduced graphene oxide (rGO)-polyaniline (PAni) conducting composite electrode for energy generation in glucose biofuel cell. Sci. Rep. **10**, 1–12 (2020). https://doi.org/10.1038/s41598-020-67253-6
53. Ramanavicius, A., German, N., Ramanaviciene, A.: Evaluation of electron transfer in electrochemical system based on immobilized gold nanoparticles and glucose oxidase. J. Electrochem. Soc. **164**, G45 (2017)
54. Malel, E., Mandler, D.: Direct electron transfer between glucose oxidase and gold nanoparticles, when size matters. ChemElectroChem **6**, 147–154 (2019). https://doi.org/10.1002/celc.201801091
55. Kwon, C.H., Ko, Y., Shin, D., Kwon, M., Park, J., Bae, W.K., Lee, S.W., Cho, J.: High-power hybrid biofuel cells using layer-by-layer assembled glucose oxidase-coated metallic cotton fibers. Nat. Commun. **9**, 4479 (2018). https://doi.org/10.1038/s41467-018-06994-5
56. Trifonov, A., Stemmer, A., Tel-Vered, R.: Enzymatic self-wiring in nanopores and its application in direct electron transfer biofuel cells. Nanoscale Adv. **1**, 347–356 (2019). https://doi.org/10.1039/c8na00177d
57. Reuillard, B., Le Goff, A., Agnès, C., Holzinger, M., Zebda, A., Gondran, C., Elouarzaki, K., Cosnier, S.: High power enzymatic biofuel cell based on naphthoquinone-mediated oxidation of glucose by glucose oxidase in a carbon nanotube 3D matrix. Phys. Chem. Chem. Phys. **15**, 4892–4896 (2013). https://doi.org/10.1039/c3cp50767j
58. Christwardana, M., Kim, K.J., Kwon, Y.: Fabrication of mediatorless/membraneless glucose/oxygen based biofuel cell using biocatalysts including glucose oxidase and laccase enzymes. Sci. Rep. **6**, 30128 (2016). https://doi.org/10.1038/srep30128
59. Courjean, O., Gao, F., Mano, N.: Deglycosylation of glucose oxidase for direct and efficient glucose electrooxidation on a glassy carbon electrode. Angew. Chemie. **121**, 6011–6013 (2009)
60. Courjean, O., Flexer, V., Prévoteau, A., Suraniti, E., Mano, N.: Effect of degree of glycosylation on charge of glucose oxidase and redox hydrogel catalytic efficiency. ChemPhysChem **11**, 2795–2797 (2010). https://doi.org/10.1002/cphc.201000178
61. Pereira, A.R., Luz, R.A.S., Lima, F.C.D.A., Crespilho, F.N.: Protein oligomerization based on Brønsted acid reaction. ACS Catal. **7**, 3082–3088 (2017)
62. Tarasevich, M.R., Yaropolov, A.I., Bogdanovskaya, V.A., Varfolomeev, S.D.: Electrocatalysis of a cathodic oxygen reduction by laccase. Bioelectrochemistry Bioenerg. **6**, 393–403 (1979). https://doi.org/10.1016/0302-4598(79)80006-9
63. Murao, S., Tanaka, N.: A new enzyme "Bilirubin Oxidase" produced by Myrothecium verrucaria MT-1. Agric. Biol. Chem. **45**, 2383–2384 (1981)
64. Tsujimura, S., Tatsumi, H., Ogawa, J., Shimizu, S., Kano, K., Ikeda, T.: Bioelectrocatalytic reduction of dioxygen to water at neutral pH using bilirubin oxidase as an enzyme and 2,2′-azinobis (3-ethylbenzothiazolin-6-sulfonate) as an electron transfer mediator. J. Electroanal. Chem. **496**, 69–75 (2001). https://doi.org/10.1016/S0022-0728(00)00239-4
65. Cracknell, J.A., McNamara, T.P., Lowe, E.D., Blanford, C.F.: Bilirubin oxidase from Myrothecium verrucaria: X-ray determination of the complete crystal structure and a rational surface

modification for enhanced electrocatalytic O_2 reduction. Dalt Trans **40**, 6668–6675 (2011). https://doi.org/10.1039/c0dt01403f
66. Dos Santos, L., Climent, V., Blanford, C.F., Armstrong, F.A.: Mechanistic studies of the "blue" Cu enzyme, bilirubin oxidase, as a highly efficient electrocatalyst for the oxygen reduction reaction. Phys. Chem. Chem. Phys. **12**, 13962–13974 (2010). https://doi.org/10.1039/c0cp00018c
67. Gupta, G., Lau, C., Rajendran, V., Colon, F., Branch, B., Ivnitski, D., Atanassov, P.: Direct electron transfer catalyzed by bilirubin oxidase for air breathing gas-diffusion electrodes. Electrochem. Commun. **13**, 247–249 (2011). https://doi.org/10.1016/j.elecom.2010.12.024
68. Quintanar, L., Stoj, C., Taylor, A.B., Hart, P.J., Kosman, D.J., Solomon, E.I.: Shall we dance? How a multicopper oxidase chooses its electron transfer partner. Acc. Chem. Res. **40**, 445–452 (2007). https://doi.org/10.1021/ar600051a
69. Shin, W., Sundaram, U.M., Cole, J.L., Zhang, H.H., Hedman, B., Hodgson, K.O., Solomon, E.I.: Chemical and spectroscopic definition of the peroxide-level intermediate in the multicopper oxidases: relevance to the catalytic mechanism of dioxygen reduction to water. J. Am. Chem. Soc. **118**, 3202–3215 (1996). https://doi.org/10.1021/ja953621e
70. Solomon, E.I., Augustine, A.J., Yoon, J.: O_2 reduction to H_2O by the multicopper oxidases. Dalt. Trans. **9226**, 3921 (2008). https://doi.org/10.1039/b800799c
71. Sekretaryova, A., Jones, S.M., Solomon, E.I.: O_2 reduction to water by high potential multicopper oxidases: contributions of the T1 copper site potential and the local environment of the trinuclear copper cluster. J. Am. Chem. Soc. **141**, 11304–11314 (2019). https://doi.org/10.1021/jacs.9b05230
72. Christenson, A., Shleev, S., Mano, N., Heller, A., Gorton, L.: Redox potentials of the blue copper sites of bilirubin oxidases. Biochim. Biophys. Acta – Bioenerg. **1757**, 1634–1641 (2006). https://doi.org/10.1016/j.bbabio.2006.08.008
73. Weigel, M.C., Tritscher, E., Lisdat, F.: Direct electrochemical conversion of bilirubin oxidase at carbon nanotube-modified glassy carbon electrodes. Electrochem. Commun. **9**, 689–693 (2007). https://doi.org/10.1016/j.elecom.2006.10.052
74. Ivnitski, D., Artyushkova, K., Atanassov, P.: Surface characterization and direct electrochemistry of redox copper centers of bilirubin oxidase from fungi Myrothecium verrucaria. Bioelectrochemistry **74**, 101–110 (2008). https://doi.org/10.1016/j.bioelechem.2008.05.003
75. Ramírez, P., Mano, N., Andreu, R., Ruzgas, T., Heller, A., Gorton, L., Shleev, S.: Direct electron transfer from graphite and functionalized gold electrodes to T1 and T2/T3 copper centers of bilirubin oxidase. Biochim. Biophys. Acta – Bioenerg. **1777**, 1364–1369 (2008). https://doi.org/10.1016/j.bbabio.2008.06.010
76. Li, Y., Zhang, J., Huang, X., Wang, T.: Construction and direct electrochemistry of orientation controlled laccase electrode. Biochem. Biophys. Res. Commun. **446**, 201–205 (2014). https://doi.org/10.1016/j.bbrc.2014.02.084
77. Macedo, L.J.A., Hassan, A., Sedenho, G.C., Crespilho, F.N.: Assessing electron transfer reactions and catalysis in multicopper oxidases with operando X-ray absorption spectroscopy. Nat. Commun. **11**, 316 (2020). https://doi.org/10.1038/s41467-019-14210-1
78. Macedo, L.J.A., Santo, A.A.E., Sedenho, G.C., Hassan, A., Iost, R.M., Feliciano, G.T., Crespilho, F.N.: Three-dimensional catalysis and the efficient bioelectrocatalysis beyond surface chemistry. J. Catal. **401**, 200–205 (2021). https://doi.org/10.1016/j.jcat.2021.07.022

Fundamentals Concepts of the Large-Scale Deposition Techniques Applied to Biodevices Manufacturing

Giovana Rosso Cagnani and Gisela Ibáñez-Redín

Abstract Large-scale fabrication techniques are the most promising pathway to produce electrochemical biosensors with low effective cost and high speeds. These techniques used as a part in roll-to-roll processing, in the discrete or integrated process mode, are attractive since they allow coating flexible substrates with solutions and dispersions of functional materials through technologies such as gravure, flexography, screen printing, inkjet, slot-die, and spray. Each of these techniques offer unique characteristics, and if combined, can allow the manufacturing of fully printed biosensors with excellent performance. In this chapter, we overviewed the particularities of large-scale techniques, such as operation parameters, ink requirements, and forces involved in the process, which are responsible for the successful deposition of the film.

Keywords Biosensor · Large scale · Roll-to-roll · Printed techniques · Coating techniques

1 Introduction

Initially, the development of electrochemical biosensors has been centered on the exploitation of analytical devices capable of detecting, quantifying, and monitoring chemical species for clinical, environmental, and industrial analyses [33]. This purpose changed facing a new trend emerging in terms of technological achievements, e.g., advances in the field of wearable sensors. Therefore, it is not enough to produce a standard sensor, it needs to be innovative and economically viable to have market significance.

Large-scale processes, such as roll-to-roll, offer the biosensor field a thriving alternative due to the versatility—is possible to manufacture films with different patterns depending on the technique used—and several advantages as minimal waste, low

G. R. Cagnani (✉)
São Carlos Institute of Chemistry, University of São Paulo (USP), São Paulo 13560-970, Brazil

G. Ibáñez-Redín
Institute of Physics, University of São Paulo (USP), São Paulo 13560-970, Brazil

F. N. Crespilho (ed.), *Advances in Bioelectrochemistry Volume 2*,
https://doi.org/10.1007/978-3-030-95270-9_3

cost, and easy manufacturing. Additionally, roll-to-roll deposition techniques can provide high throughput fabrication; besides the materials which are processed in solution and directly deposited on flexible substrates, it is possible to manipulate various materials such as nanoparticles, semiconductor polymers and carbon materials at high speed. There is a wide variety of roll-to-roll deposition techniques, such as spray, inkjet, flexography, slot-die, gravure, and screen printing, with specific characteristics that can be used to obtain a fully printed biosensor. The choice between one and another depends on the unit operations (drying and UV curing), deposition characteristics (thickness, number of layers, pattern), and ink properties (viscosity and surface tension) [6]. To use spray technique, for example, the ink must have low viscosity to produce high print resolutions and control over material thickness [24]. The inkjet printing pre-establishes that inks are compounds with small and enough dispersed particles to not clog the inkjet nozzle [11]. While spray and inkjet operate at low speed, flexographic, and gravure printing can work at high speeds [3, 34] using inks with viscosity ranging from 50–500 mPa.s. Slot-die coating can also work at high speeds and with viscous inks; on the other hand, there is no possibility of obtaining a specific printed pattern [25] as the other techniques listed.

Generally, to be successful in the manufacture of electrochemical biosensors by any deposition technique, the technical aspects of production must be aligned with the particularities of the ink or vice versa. The ink composition must contain physical properties that lead to a satisfactory application. This implies that it should be possible to spread the ink over the surface of the substrate chosen for the biosensor (obtaining a regular film); that the wet film should be able to remain where it is deposited (film must not leak); and, that the drying of the film should be uniform [15]. To clarify, the decision of what technique to use depends on the number of layers deposited, what part of the biosensor will be produced (reference, working, or counter electrode), and which ink will be deposited—although some inks can be modified to be deposited by different techniques. However, the particularity of the ink used in the deposition process depends on the type of electrochemical biosensor being manufactured.

Enzyme biosensors require deposition techniques with low or no shear stress to avoid the loss of enzymatic activity [22]. As for the preparation of inks, it is possible to make different formulations without harming the biosensor. Immunosensors require preparation steps for active film deposition. Is common for antibodies to be deposited onto films that have the function of anchoring and organizing biomolecules to later target recognition. Inks that contain antibodies in the composition are more aqueous and need to be manipulated in a way that dewetting does not occur [16] when they are deposited. Biosensors that use nucleic acid as a biorecognition element, in general, have the same requirements as immunosensors, with only some variations in the previous layers of the active element.

In this chapter, we will present the fundamentals of the techniques compatible with roll-to-roll processing, which is used the most to produce electrochemical biosensors. We will begin introducing the roll-to-roll processing and classifying deposition techniques as printing or coating techniques. Subsequently, we will discuss the key manufacturing methods of electrochemical biosensors, emphasizing how deposition physics leads to constraints on ink design, the deposition characteristics, and the print speed of each technique addressed.

2 Roll-to-Roll Processing

Roll-to-Roll is a process in which the movement of the substrate through two or more rolls forms a thin film of ink, therefrom building continuously different architectures [1]. From a fluid mechanics perspective, in the roll-to-roll process, the fluid flows through the existing nip between a pair of rotating rolls or through depositing tools that control the thickness and uniformity of the film [20]. This process involves physical phenomena associated with the air–liquid interface, static and dynamic wetting lines, and the rheology of fluids, that in most cases do not follow Newton's law of viscosity [7]. However, how these phenomena act in the film formation is directly related to the technique used, as well as the forces that govern the process, which are: (i) viscosity—drag oriented in the flow direction; (ii) capillarity—the action of surface tension at the interface of the coating surface; (iii) Elastic force—resistance to deformation in the solid phase; and, (iv) external force—applied force to move the rollers (substrate rotation) [7].

To ensure satisfactory deposition, the physical factors must be aligned with the process parameters, considering that a lot of steps will be necessary to produce a fully printed biosensor using the roll-to-roll process. First, the substrate is unrolled from the initial bobine, passes through the deposition tools (where physical phenomena act) and, at the end of the process; it is rewound in the external bobine. In addition to deposition operations, there are other processes such as substrate treatment, drying, UV curing, etc. Each of these steps requires individual adjusts such as scrolling speeds, tension, and alignment of the substrate, making the planning of fully integrated process often difficult. Alternatively, a discrete process is used, thus each layer is processed individually, and the roll of material is transferred to the next equipment. Figure 1 shows a discrete (Fig. 1a) and integrated (Fig. 1b) roll-to-roll process [14].

Normally, discrete processes are used to evaluate the processing parameters and drying mechanisms for different deposition techniques and types of ink. These characteristics enable the optimization of the processes and materials involved, until they reach sufficient maturity to be applied in an integrated production line.

3 Deposition Techniques Compatible with Roll-to-Roll Processing Used in the Manufacturing of Biosensors

This section will present the deposition techniques, compatible with roll-to-roll processing, most used for biosensors manufacturing. However, before presenting them it will be necessary to clarify some concepts about printing and coating.

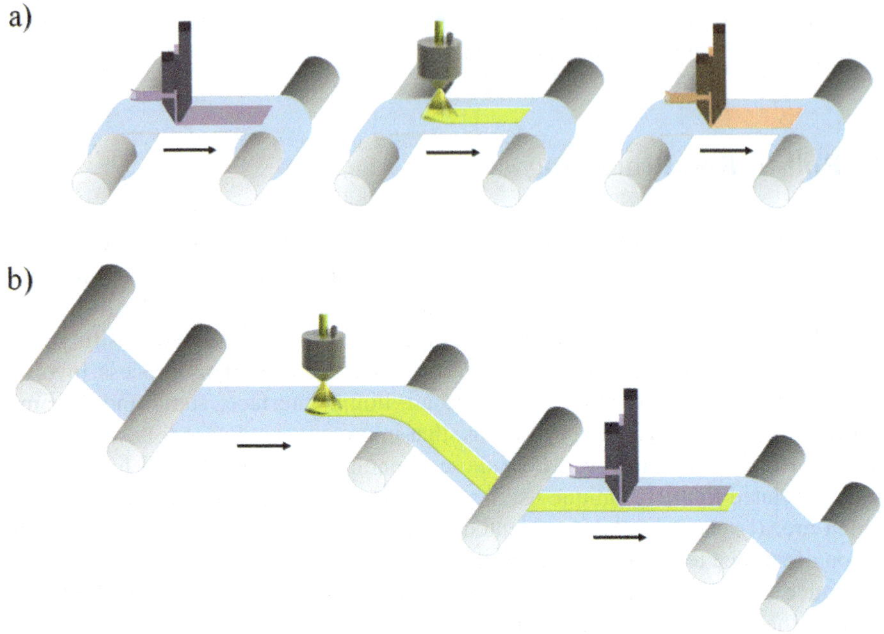

Fig. 1 Roll-to-roll process in **a** discrete and **b** integrate mode

All thin film production processes conducted in solution onto a substrate can be classified as a deposition process. The deposition processes are divided into two main categories: printing and coating. Printing techniques are characterized by the close contact between the printing tool and the substrate and are used for depositing defined patterns at different scales [14]. Rotogravure, flexography, and screen printing are examples of printing techniques. In coating techniques, there is no contact between the deposition tool and the substrate, so it is impossible to produce patterns as drawings, signs, and logos. Coating processes are often used in the deposition of continuous films [14]. Slot-die is an example of the coating techniques. However, some techniques are classified into both groups. The spray technique, which is a coating technique, allows the formation of patterns by using molds fixed to the substrate containing the desired pattern. Once the concepts of coating and printing have been defined, the deposition techniques compatible with roll-to-roll processing used for manufacturing biosensors will be presented.

3.1 Printing Techniques

3.1.1 Screen Printing

Screen printing is a process characterized by pressing the ink through a screen by a squeegee or puller. This action transfers the pattern that was previously engraved in the screen printing matrix to the substrate positioned below the printing system [19]. The screen printing matrix, stretched on a wooden, or aluminum frame (screen) has a defined mesh, i.e., a number of threads per square centimeter that correspond to the linescreen. The definition of points of the pattern to be engraved is calculated from the density of threads of the matrix that is made of polyester or nylon fabrics. The preparation of the screen and engraving of the pattern occurs through the photosensitivity process. Initially, the matrix is entirely covered with a photosensitive emulsion in a dark place. Afterward, the shadow mask with the printing pattern design is placed over the matrix and these set is placed under the light source, as indicated in Fig. 2. The dark regions of the shadow mask block the light passage (Fig. 2a) and correspond to the place that the ink leaks through the fabric weft (Fig. 2b). The other regions (regions without the print pattern) are waterproofed by curing the resin and hardening the photosensitive emulsion due to light exposure.

During the printing process, the engraved screen is placed over the substrate (Fig. 3a), without touching it, maintaining a minimum distance (D) (Fig. 3b). Excess ink is spilled in a side of the matrix and then is scattered across the screen, filling the print pattern, as shown in Fig. 3a. Due to the pressure applied with a squeegee on the screen, the ink leaks between the uncovered threads of the matrix. The elasticity of the fabric that the screen is made of allows temporary contact of the matrix with the substrate [19]. After the application of pressure ceases, the screen returns to the original state and the print pattern is transferred, as indicated in Fig. 3b.

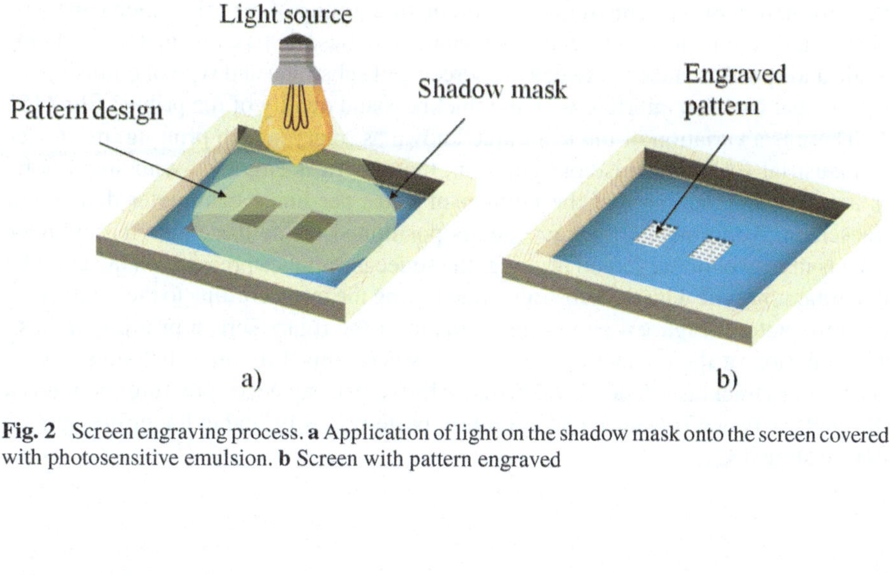

Fig. 2 Screen engraving process. **a** Application of light on the shadow mask onto the screen covered with photosensitive emulsion. **b** Screen with pattern engraved

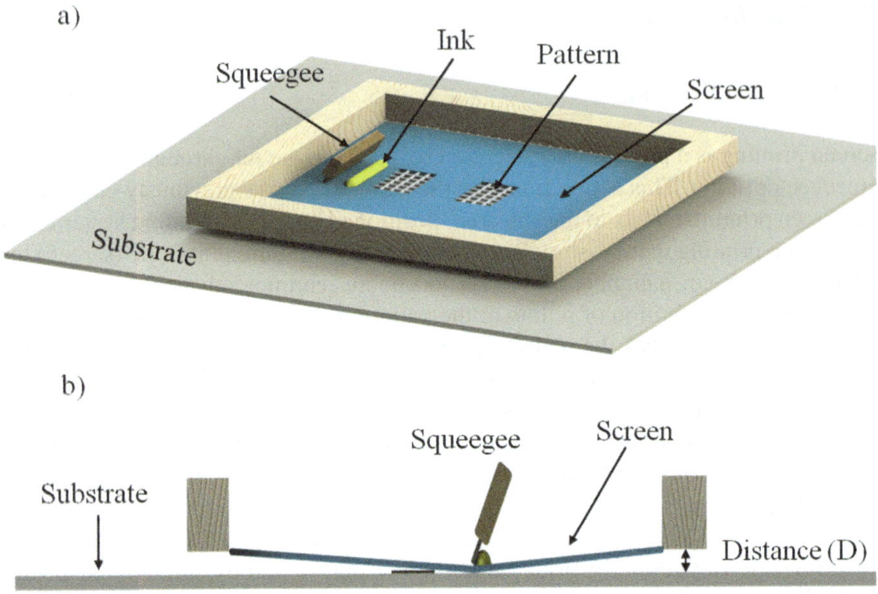

Fig. 3 Screen printing process. **a** Position of the screen on the substrate and application of the ink. **b** The squeegee movement and printing of the engraved pattern

The screen printing technique uses inks containing high viscosity, usually in the form of pastes, with the finality of preventing leakage of the ink through the matrix as it is spilled onto the screen. Due to the characteristics of the inks, the printed film is relatively thick before drying (>10 μm) [32] and, for this reason, it is recommended to use solvents with low evaporation rate, allowing accommodation of the printed film without marks of matrix threads. Solvents with high evaporation rate cause poor formation of the printed film and premature drying of the ink, which can cause clogging of the matrix weft. Substrate characteristics, material of matrix, scraping angle and speed, distance between the screen and substrate, and type of emulsion are factors that can also interfere with the thickness and quality of the printed film [15].

There is a variation of the technique known as rotary screen printing, most used in industrial scales. In this configuration, the pattern is engraved with small holes in a nickel cylinder, so that the number of holes per linear centimeter defines the linescreen. Additionally, the squeegee is positioned inside the engraving cylinder. As it is in conventional screen printing, the squeegee applies pressure on the cylinder that rotates in sync with the substrate, transferring the ink according to the established printing pattern. Figure 4 shows a schematic of the rotary screen printing process. The duration of the contact between them is determined by the substrate scrolling speed. Film thickness depends on cylinder linescreen, squeegee pressure (defined by the angle between squeegee and cylinder), paste viscosity, and cylinder pressure on the substrate [32].

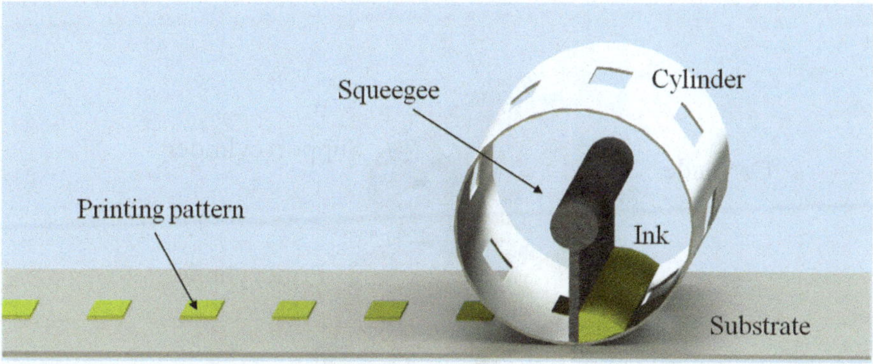

Fig. 4 Scheme of the rotary screen printing process. Printing cylinder cross-section and position of the squeegee, the ink, and the printed pattern

The advantages of the screen printing process—whether conventional or rotary—are low operating cost, ease of changing the print pattern and compatibility with roll-to-roll production. The difficulty in obtaining thin films (less than 1 μm) and the need to use inks with high viscosity are some limitations of this technique. Due to ink requirements, screen printing is commonly used only for producing some parts of the biosensors, such as working, reference, and counter electrodes [5, 12].

3.1.2 Gravure Printing

The simplest gravure printing process consists in a system of two cylinders, one of support and other of print, as shown in Fig. 5. The print cylinder, which contains the printing pattern, is partially inserted into the ink bath to be refilled continuously. The excess of ink is removed using a doctor blade before coming into contact with the substrate, leaving only the ink of the cavities of engraved pattern [27]. The print cylinder is tensioned against the support cylinder, that controls the substrate, and then, the pattern is transferred by the contact between the substrate and the print cylinder [29].

Often, the support cylinder is encased with rubber to ensure good contact between the substrate and the print cylinder. The print cylinder receives electrostatic nickel and copper baths over which the printing pattern is engraved by electromechanical, chemical, and laser techniques. Posteriorly, it is covered with a chrome electrolytic bath to ensure the cylinder's durability against the scraping process. The printing pattern is defined as the percentage of the total cavity volume, which can range between values from 0 to 100% and represents the tone of the print cylinder. Several parameters define the printing quality and thickness of the film deposited in the gravure process. These factors are divided into three groups: printing parameters, characteristics of the inks, and cylinder engraving. The printing parameters involve the rotation speed of the print and the support cylinders, contact pressure between the

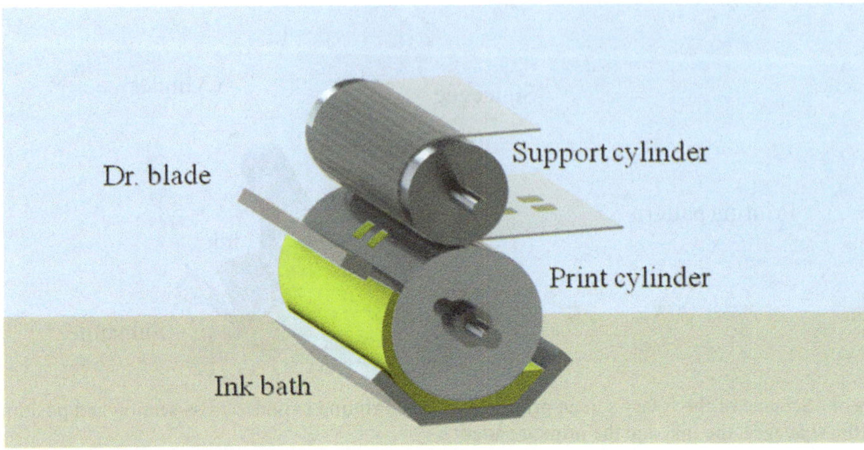

Fig. 5 Scheme of the direct gravure printing process showing the support cylinder, print cylinder, doctor blade, and ink bath

cylinders, and doctor blade inclination angle. The characteristics required for inks include properties such as viscosity—responsible for keeping the ink in the pore; surface tension—which ensure the entry of ink into the cavity; solids concentration—in relation to the dry thickness; and solvent type—evaporation rate that defines the film uniformity [10].

In the manufacture of biosensors, the gravure process can be used both to print electrodes and to print functional materials or biomolecules. However, the desired characteristics for each printed layer must be evaluated, because for biomolecules the spatial orientation is important for the biorecognition process with the target.

3.1.3 Flexography Printing

The flexographic printing process begins by supplying the ink to the anilox cylinder, a ceramic cylinder with microcavities on the surface. These microcavities control spacing and specific ink volumes per area [27]. The anilox cylinder is continuously filled from the contact with a supply cylinder or from partial immersion in an ink bath. Excess of ink in the anilox is removed using a doctor blade that ensures good control of the transferred ink volume. Afterward, the ink from the anilox cylinder is transferred to the soft printing plate cylinder, which is the printing pattern itself, made of rubber or an embossed photopolymer and fixed with a special double-sided tape. Both the soft printing plate and the tape have defined variations in thickness and hardness, indicated according to the application. The printing of the pattern occurs through direct contact between the printing cylinder and the substrate, controlled by the support cylinder [21]. Is possible to observe in Fig. 6a how the cylinders are

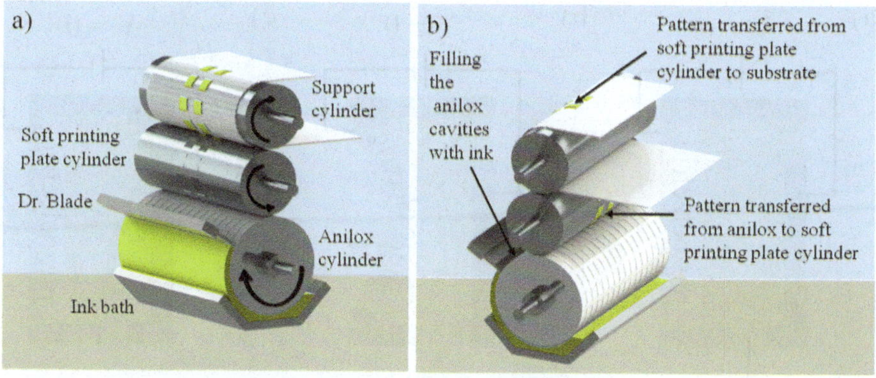

Fig. 6 Scheme of flexography printing process. **a** Position of anilox, soft printing plate, and support cylinders. **b** Transferring of the ink to the substrate

positioned and how the ink is transferred to the substrate in flexography printing process, Fig. 6b.

For the film printed by flexography to present uniformity, the inks used must have low viscosity. This allows the ink to quickly flow from the print cylinder to the substrate. Inks with high viscosity prevent the correct filling of the printing pattern and the proper transfer to the substrate due to the surface tension effect created on the printing pattern walls, causing problems such as clogging and loss of print detail.

As with gravure printing, the flexographic printing process can be used to print several layers of biosensor as the electrodes, functional materials, and biomolecules. Although, in flexography printing, the more intense shear stress between the blade and the soft printing plate cylinder—or between cylinders—can compromise the stability of biomolecules.

3.1.4 Inkjet Printing

The inkjet printing process is carried by the deposition of small individual ink drops, so that each drop corresponds to one pixel of the pattern printed on the substrate after evaporation of the solvent [13]. Inkjet printers can operate in two modes: continuous and drop-on-demand (DOD) mode [31], the latter being the most used widely.

In DOD mode, the ink drop formation takes place in two ways: ink compression through the print nozzle (piezoelectric) or by heating the ink, both demonstrated in Fig. 7. In the piezoelectric actuation system (Fig. 7a), the application of a potential in the piezoelectric element in contact with the printer nozzle promotes a volume displacement (Fig. 7b) and the ejection of one ink drop (Fig. 7c) [18, 31]. On the thermal actuation system (Fig. 7d), drop ejection occurs by the formation of gas bubbles due to evaporation of the solvent in contact with the heating element (Fig. 7e), increasing internal pressure and expelling a volume of ink through the printing nozzle (Fig. 7f) [13, 18]. Due to the drop formation process, the inks used in the inkjet

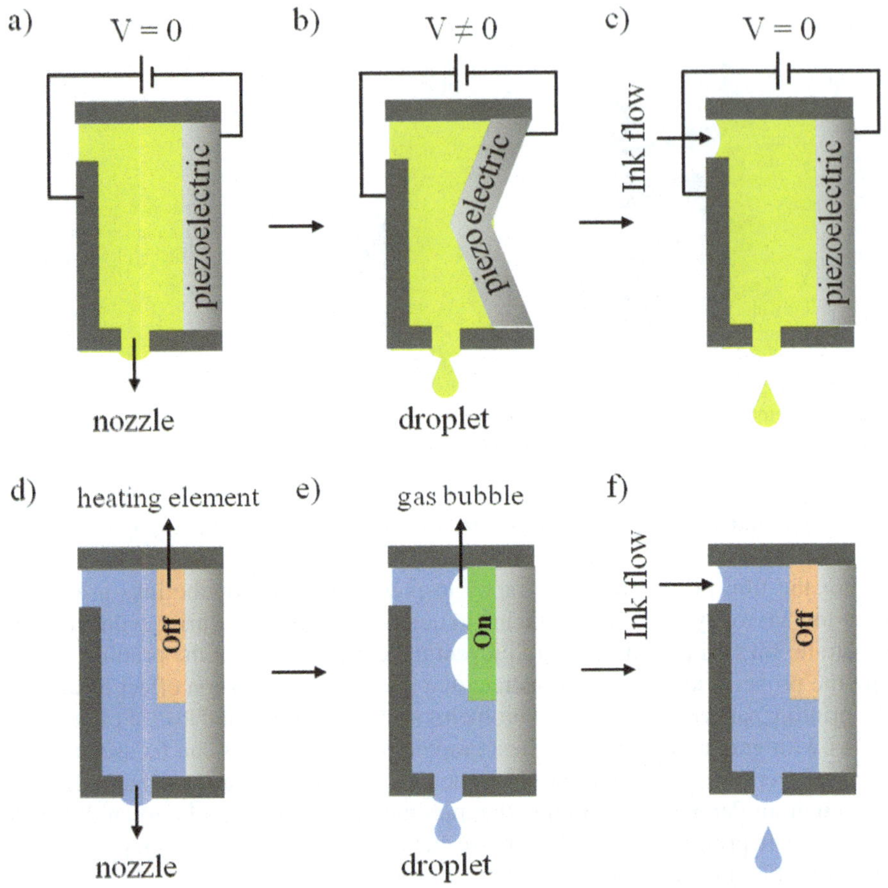

Fig. 7 Droplet ejection in **a** thermally and **d** piezoelectric system actuation of inkjet printers operated in DOD mode. **b** Contraction of piezoelectric element and the formation of the droplet. **c** Droplet ejection in piezoelectric system actuation **e** Gas bubble formation by the actuation of heating element. **f** Droplet ejection in thermal system actuation. Adapted from [18]

printing process must have low viscosity and high surface tension to easily generate the flow of droplets.

The dry thickness of films printed using the inkjet technique can be calculated from the number of drops delivered per area, the individual drop volume and the concentration of solid materials in the ink [14], by the following relationship:

$$d = N_d \times V_d \frac{C}{\rho} \tag{1}$$

where N_d is the number of droplets delivered per area (cm^2), V_d is the volume of the droplets (cm^3), C is the concentration of the ink (g cm^{-3}), and ρ is the density of the ink (g cm^{-3}).

In contrast to other printing techniques, where there is a pre-engraved pattern in the deposition tool, on inkjet printing this pattern is easily modified in a computer program. Thus, in terms of design, is acceptable that a biosensor can be completely manufactured by inkjet printing. However, factors such as organization of biomolecules, loss of activity (for enzymes), and production speed (inkjet printing operates at low speed) must be carefully analyzed.

3.2 *Coating Techniques*

3.2.1 Spray

The use of the spray nozzle characterizes the film deposition using spray coating technique. In this process, the film deposition is made by pressurizing the hose that supplies ink to the spray nozzle, in this way the controlled entry of air into the hose creates small droplets of ink which are launched onto the substrate positioned below the nozzle, as indicated in Fig. 8. The kinetic energy of the droplets helps them spread over the substrate upon impact, obligating the formation of a uniform layer [27].

Variations in the nozzle control parameters such as air entrance, ink volume, a mixture of air/ink inside the nozzle, and nozzle opening in the exit-atomized fluid are responsible for producing ink deposits with different characteristics [30]. Furthermore, different nozzle formats as well as conical, plane, and directional (focused on a point)—illustrated in Fig. 9—also modify the structure of the deposited films. Therefore, the film thickness manufactured using the spray technique is determined from the distance between the spray nozzle and the substrate, the coating speed and the number of sprayed layers [28].

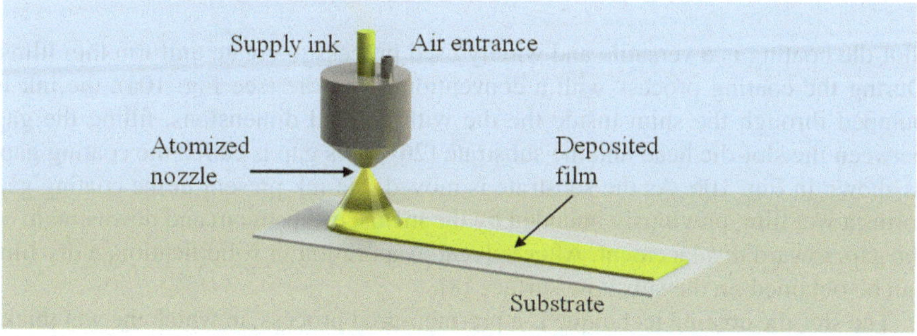

Fig. 8 Scheme of spray coating process with an indication of air entrance, ink supply, and atomized nozzle

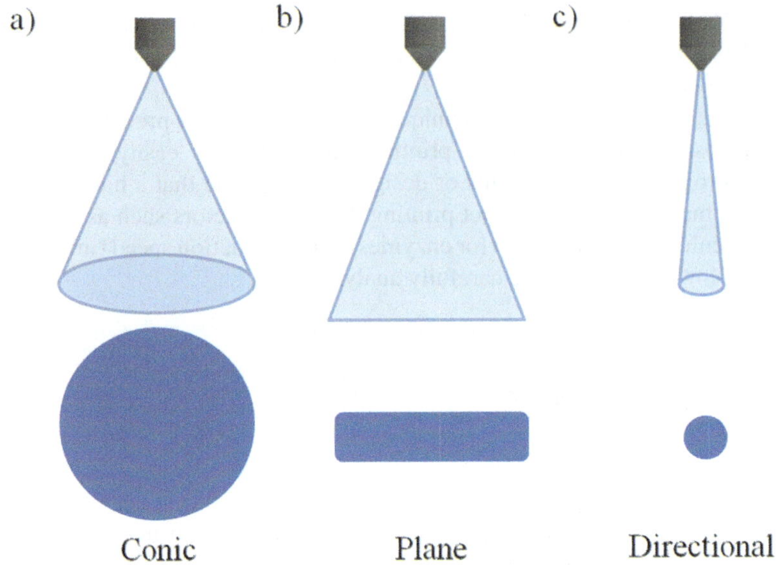

Fig. 9 Spray nozzle formats commonly used: **a** Conic, **b** plane, and **c** directional

Another important parameter is the drying temperature of the deposited film. The drying temperature must be chosen according to the specifications of the solvent or the solvent mixture used in the ink [2]. High temperatures promote rapid evaporation of the solvent, which can cause defects such as the formation of bubbles, delamination and drying of the droplets preventing the formation of uniform film [9]. The disadvantage of using spray coating is a limitation due to the viscosity of the inks. High viscous inks cause spray nozzle clogging and for this reason the spray technique is selected to functionalize the biosensor work electrode.

3.2.2 Slot-Die Coating

Slot-die coating is a versatile and widely used process to create uniform thin films. During the coating process with a conventional slot-die (see Fig. 10a), the ink is pumped through the shim inside the die with defined dimensions, filling the gap between the slot-die head and the substrate [26]. This gap is called the coating gap, as shown in Fig. 10b. As the substrate is moved, the ink present in the coating gap forms a wet film, previously bounded by the meniscus upstream and downstream of the gap, toward displacement. After solvent evaporation or solidification, a dry film can be obtained on the substrate surface [8].

The slot-die coating technique is a pre-measured process, in which the wet thickness can be pre-set and precisely controlled by adjusting the processing parameters, as shown in Eq. 2 [14]. Already, the approximate dry film thickness is obtained from the product between Eq. 2 and the relation between the concentration of solids and

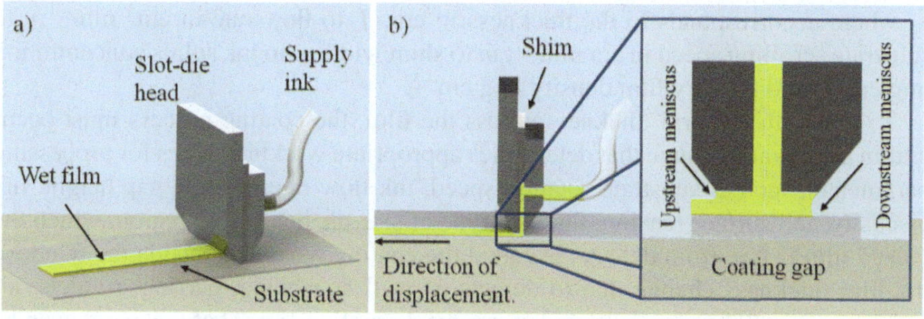

Fig. 10 Slot-die coating technique. **a** Conventional slot-die coating with single deposition head. **b** Cross-section of slot-die head showing the shim and expanded image of coating gap

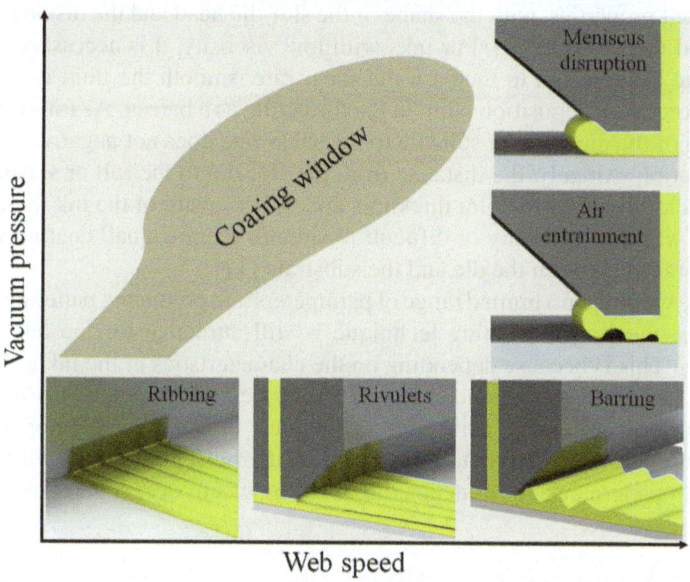

Fig. 11 Coating window and regions outside the window where defects as ribbing, rivulets, barring, air entrainment in the ink bed, and meniscus disruption occur due to the combinations of process parameters established outside the operating range. Adapted from [23]

the density of the ink, Eq. 3 [14].

$$d = \frac{f}{v \times w} \tag{2}$$

$$d = \frac{f}{v \times w} \times \frac{c}{\rho} \tag{3}$$

where d corresponds to the thickness in cm, f to flow rate in $cm^3\ min^{-1}$,v to substrate scrolling speed in cm min^{-1}, w to shim width, c to ink solids concentration in g cm^{-3}, and ρ to dry film density in g cm^{-3}.

To maintain uniform thickness across the film, the coating process must occur within an operating range that determines appropriate working values for processing parameters, such as substrate rotation speed, ink flow rate, coating gap height, ink viscosity, and surface tension, for each type of slot-die head. This region, which the coated film is free from defects, is called the coating window. Outside the window, the film thickness change due to defects as well as ribbing, barring, rivulets, air entrainment into the meniscus, and meniscus disruption, e.g. There may not even be a continuous layer covering the substrate [23]. Figure 11 shows the defect-free region for a conventional slot-die die and the coating defects outside the coating window.

The region of the coating window is related, besides the processing parameters and physical properties, with the shape of the slot-die head and the distance between the die and the substrate [4]. For inks with low viscosity, it is necessary to modify the slot-die head design to increase the shear rate, smooth the fluid in the coating gap, and force film deposition without the die acting as a barrier. As for viscous inks, the design of the die must be suitable to the shear rate does not negatively affect the film formation. Already, the distance from the slot-die to the roll or substrate is, in general, determined by the film thickness and the viscosity of the ink to be applied. Materials with low viscosity or difficult to smooth require small coating gaps, that is, less distance between the die and the substrate [17].

Despite working in a limited range of parameters and producing patternless continuous films, the slot-die coating technique is still attractive for the fabrication of biosensors. This is because depending on the characteristics of the ink used, mainly related to its viscosity, the adjustment of the processing parameters is more easily done than in other coating or printing techniques. In addition, it is possible to use this technique for the complete fabrication of the biosensor, safeguard the use of masks with pattern design for depositing working electrodes, counter electrode and reference electrode.

4 Conclusion and Outlook

This chapter outlines some roll-to-roll compatible large-scale manufacturing techniques with application in the development of electrochemical biosensors. Screen printing, gravure, flexography, and inkjet were the printing techniques presented in this chapter. Whereas, spray and slot-die were the coating techniques cited. For each technique mentioned, the physical properties and the process parameters that interfere in the deposition were highlighted. In roll-to-roll processing, the deposition steps demands more attention, as it directly influences in the building or functionalization of biosensors electrode. Is at this stage that the coating or printing techniques are selected, and the choice between one and the other depends on the ink properties, as well as on the characteristics required for the biosensor.

In general, techniques that require the use of viscous inks or allow the formation of printed patterns are used for the fabrication of the biosensor electrodes. Techniques that use inks with low viscosity and produce continuous films are used to modify the surface of the working electrode. Depending on the type of biosensor being manufactured, techniques with high shear stress should be discarded, as they can interfere in the properties of the active element. Thus, before any initiative to produce a biosensor using large-scale deposition techniques, it is correct evaluated the processes in a discrete mode, i.e., separately. This ensures the control of the variables involved in the deposition systems and contributes to the development of an integrated process for the production of fully printed devices using techniques compatible with roll-to-roll processing. Large-scale deposition processes have emerged as a promising trend to reduce the cost of manufacturing sensors and biosensors.

Acknowledgements Giovana Rosso Cagnani gratefully acknowledges the Postdoctoral grant by MeDiCo Network CAPES-Brazil (23038.003012/2020-16) and Gisela Ibáñez-Redín acknowledges the technical training grant by FAPESP (2020/14906-5).

References

1. Abbel, R., Galagan, Y., Groen, P.: Roll-to-roll fabrication of solution processed electronics. Adv. Eng. Mater. **20**(8), 1701190 (2018)
2. Aziz, F., Ismail, A.F.: Spray coating methods for polymer solar cells fabrication: A review. Mater. Sci. Semicond. Process. **39**, 416–425 (2015)
3. Bariya, M., et al.: Roll-to-roll gravure printed electrochemical sensors for wearable and medical devices. ACS Nano **12**(7), 6978–6987 (2018)
4. Carvalho, M.S., Kheshgi, H.S.: Low-flow limit in slot coating: theory and experiments. AIChE J. **46**(10), 1907–1917 (2000)
5. Choi, H., et al.: UV-curable silver electrode for screen-printed thermoelectric generator. Adv. Func. Mater. **29**(20), 1901505 (2019)
6. Cohen, E.D., Gutoff, E.B.: Modern coating and drying technology. University of Texas Press (1992)
7. Coyle, D.: Roll Coating. In: Cohen, E.D., Gutoff, E.B. (eds.) Modern Coating and Drying Technology, p. 336. Wiley, New York (1992)
8. Ding, X., Liu, J., Harris, T.A.L.: A review of the operating limits in slot die coating processes. AIChE J. **62**(7), 2508–2524 (2016)
9. Girotto, C., et al.: High-performance organic solar cells with spray-coated hole-transport and active layers. Adv. Func. Mater. **21**(1), 64–72 (2011)
10. Grau, G., et al.: Gravure-printed electronics: recent progress in tooling development, understanding of printing physics, and realization of printed devices. Flex. Print. Electron. **1**(2), 23002 (2016)
11. Huang, Q., Zhu, Y.: Printing conductive nanomaterials for flexible and stretchable electronics: a review of materials, processes, and applications. Adv. Mater. Technol. **4**(5), 1800546 (2019)
12. Josypčuk, B., Langmaier, J., Tvorynska, S.: Screen-printed amalgam electrodes. Sens. Actuators B: Chem. **347**, 130583 (2021)
13. Komuro, N., et al.: Inkjet printed (bio) chemical sensing devices. Anal. Bioanal. Chem. **405**(17), 5785–5805 (2013)
14. Krebs, F.C.: Fabrication and processing of polymer solar cells: a review of printing and coating techniques. Sol. Energy Mater. Sol. Cells **93**(4), 394–412 (2009)

15. Leach, R.: *The printing ink manual*. Springer Science & Business Media (2012).
16. Leroy, F., et al.: How to control solid state dewetting: a short review. Surf. Sci. Rep. **71**(2), 391–409 (2016)
17. Lippert, H.G.: Slot-die coating for low viscosity fluids. In: Tracton, A.A. (ed.) Coatings technology: fundamentals, testing, and processing techniques, 3rd edn., p. 370. CRC Press, Boca Raton, FL (2007)
18. Mattana, G., et al.: Inkjet-printing: a new fabrication technology for organic transistors. Adv. Mater. Technol. **2**(10), 1700063 (2017)
19. Mc Sweeney, T.B.: Screen printing. In Tracton, A. A. (ed.) *Coatings Technology: Fundamentals, Testing, and Processing Techniques*, 3rd edn, p. 370. CRC press (2007).
20. Morse, J.D.: Nanofabrication Technologies for Roll-to-Roll Processing. NIST-NNN Workshop, Boston, MA (2011)
21. Neumann, R.: Flexography. In: Tracton, A.A. (ed.) Coatings Technology: Fundamentals, Testing, and Processing Techniques, 3rd edn., p. 370. CRC Press, Boca Raton (2007)
22. Pavinatto, F.J., Paschoal, C.W.A., Arias, A.C.: Printed and flexible biosensor for antioxidants using interdigitated ink-jetted electrodes and gravure-deposited active layer. Biosens. Bioelectron. **67**, 553–559 (2015). Doi: https://doi.org/10.1016/j.bios.2014.09.039
23. Sartor, L.: Slot coating: fluid mechanics and die design. University of Minnesota (1990)
24. Secor, E.B.: Principles of aerosol jet printing. Flex. Print. Electron. **3**(3), 35002 (2018)
25. Seong, J., et al.: Practical design guidelines for the development of high-precision roll-to-roll slot-die coating equipment and the process. IEEE Trans. Compon. Packag. Manuf. Technol. **6**(11), 1677–1686 (2016)
26. Søndergaard, R., et al.: Roll-to-roll fabrication of polymer solar cells. Mater. Today **15**(1–2), 36–49 (2012)
27. Søndergaard, R.R., Hösel, M., Krebs, F.C.: Roll-to-Roll fabrication of large area functional organic materials. J. Polym. Sci. Part B: Polym. Phys. **51**(1), 16–34 (2013). https://doi.org/10.1002/polb.23192
28. Steirer, K.X., et al.: Ultrasonic spray deposition for production of organic solar cells. Sol. Energy Mater. Sol. Cells **93**(4), 447–453 (2009)
29. Subramanian, V., et al.: *Printing Techniques for Thin-Film Electronics*. Wiley-VCH Verlag GmbH & Co. KGaA, Weinheim, Germany (2012).
30. Susanna, G., et al.: Airbrush spray-coating of polymer bulk-heterojunction solar cells. Sol. Energy Mater. Sol. Cells **95**(7), 1775–1778 (2011)
31. Tortorich, R.P., Choi, J.-W.: Inkjet printing of carbon nanotubes. Nanomaterials **3**(3), 453–468 (2013)
32. Tracton, A.A.: *Coatings Technology-Fundamentals, Testing, and Processing techniques*. United States of America: Taylor & Francis Group, LLC (2007).
33. Zavanelli, N., Kim, J., Yeo, W.H.: Recent advances in high-throughput nanomaterial manufacturing for hybrid flexible bioelectronics. Materials **14**(11) (2021). Doi: https://doi.org/10.3390/ma14112973
34. Zhong, Z.W., et al.: Parametric investigation of flexographic printing processes for R2R printed electronics. Mater. Manuf. Processes **35**(5), 564–571 (2020)

Volume-Confined Biomolecules for Application in Biocatalysis

Rodrigo M. Iost

Abstract Enzymes can catalyze innumerous metabolic process in living organisms. They are involved in all chemical pathways that are essential for life and are part of a complex network of biological functions that realize efficient and selective biochemical reactions. An important characteristic of such natural catalysts is that their activity is usually regulated by their spatial organization, forming macromolecular structures, eventually being cooperatively part of multistep or cascade chemical reactions. Advanced scientific studies in this field aims to future technological applications, such as the development new generation of artificial environments to carry out a desirable chemical synthesis or even for target drug release applications. Many efforts to transfer such biomolecules to a non-native environment require that they are either bound or adsorbed to a membrane, or even physically confined in to a volume defined by an artificial compartment, at micro or nanoscale level. However, the creation of an artificial environment, able to perform desirable chemical reactions, is not a simple task. One of the reasons for that is the difficulty to obtain a synthetic media comparable to the conditions of biological medium. Herein, we will discuss some of the main strategies used for the spatial confinement of biomolecules in a specific volume for application in biocatalysis. For this purpose, we will focus on the use of polymer vesicles, liposomes, and virus-like particles as the main strategies used for micro- and nanoconfinement of enzymes.

Keywords Biocatalysis · Biomolecules · Immobilization · Volume-Confined

1 Introduction

1.1 Volume-Confined Biomolecules

The biological activity of biomolecules in diluted conditions can be different from their activity in crowding conditions [1]. The effect of crowding condition alter the

---

R. M. Iost (✉)
Instituto de Química de São Carlos, Universidade de São Paulo, São Carlos, SP 13560-970, Brazil
e-mail: rodrigoiost@usp.br

F. N. Crespilho (ed.), *Advances in Bioelectrochemistry Volume 2*,
https://doi.org/10.1007/978-3-030-95270-9_4

structure and functions of enzymes and can be exemplified when enzymes undergo its biological activity inside living cells [2]. Cells contains a relative high concentration of macromolecules and crowding conditions that affects the spatial confinement of enzymes [3]. This is basically an important consideration when artificial micro- and nanoreactors are designed [4]. The volume-confinement of metabolic enzymes could improve their catalytic efficiency by favoring consecutive reactions that are placed in close proximity to each other, reducing the diffusion distance of metabolites and performing a maximum efficiency of an enzymatic reaction [5]. For this purpose, an overaching concept for engineering synthetic compartments from polymeric materials and biomaterials has been stablished for numerous practical applications, such as in biosynthesis, gene therapy, tissue engineering, or medicine. When medical applications are desirable, studies of biocompatibility [6], stability in vitro and in vivo [7], and biodegradability mechanisms [8] are also necessary. For example, major efforts are focused on the use of synthetic polymers vesicles (polymersomes), liposomes, proteins, or virus-like particles for fabrication of the capsules [9].

Polymersomes are hollow spheres structures inspired by natural phospholipids found in living cells that comprises water in its core and are surrounded by a block copolymer membrane [10]. If properly designed, polymersomes can achieve controlled drug release through stimuli-responsive chemistry [11], e.g., by incorporation of proteins and other protocols for surface functionalization with specific ligands [12] or even used as emissive fluorophores to give rise a localized optical signal for tumor issue analysis [13]. Many studies have been reported in the literature aiming the mimicking of natural compartments for biological reactions [14]. One of the fundamental problems is the proper choice of the material used for the compartmentalization of biological molecules, such as by the use of layer-by-layer deposition [15], sol–gel chemistry [16], or using capillary microfluidic devices for and efficient encapsulation of the proteins in double emulsion drops [17]. Due to the versatile properties of polymersomes and their ability to compartmentalize aqueous and different organic solvents, their applications can be extended to design chemical reactions in micro and nanoreactors, e.g., to produce adenosine triphosphate (ATP) [18]. The use of polymersomes as encapsulating agents of enzymes can be then advantageous to control the release of some specific products of reaction. The entrapment of enzymes within polymersomes can be used as active micro- or nanoreactors and the integration of selective membrane channel proteins can be incorporated to provide selective mass transfer for specific products of reaction [12]. As an example, polymersomes can undergoes biocatalysis in biphasic reaction systems at which enzymes are present in aqueous phase and the reaction products are transferred to a specific organic solvent outside the membrane [19]. Another important aspect for the production of polymersomes is the high level of control of the enzymes in separate domains within polymersomes to carry out cascade reactions.

The compartmentalization of enzymes are an important step to mimic the coupling biological process in nature, such as by the incorporation of enzymes in polymersomes. For example, the investigation of the structural stability formed by a diblock copolymer polymersome with selective incorporation of *Candida antarctica lipase B* (CALB), horseradish peroxidase (HRP), and glucose oxidase (GOx) [20] was carry out using transmission electron microscopy (TEM), with a clear evidence that the incorporation of the enzymes did not disrupt the structure of the nanoreactor [20]. The diameter of the polymersomes was around 50–1100 nm (average diameter of 517 nm). The enzyme activity in to polymersomes as well as the confirmation of the incorporation in to the polymersomes was then monitored by fluorescence spectroscopy. The enzyme CALB labeled with Alexa Fluor 48 was incorporated in to the membrane of the polymersome. Both soluble and aggregates was located in the membrane and in the internal water pool of the polymersome and independently studied by switching the excitation wavelengths from 488 to 633 nm, respectively, and highlights the positional assembly of the enzymes in different regions of the polymersome using lyophilization approach [20]. The positional assembly was also extended for the incorporation of the enzymes GOx and HRP. HRP was incorporated in to the hydrophilic domains of the membrane of the polymersome and GOx was entrapped in to the water pool of the polymersome. The coupling reaction between GOx/HRP was firstly monitored spectroscopically. In this experimental approach, the reaction pathway from the outside of the polymersome through the enzyme GOx in the inner pool reaches the enzyme HRP located in to the membrane. The reaction pathway was followed by the presence of 2, 2′-azinobis(3-ethyl-benzothiazoline-6-sulphonic acid) (ABTS). The reaction pathway proceeds smoothly through the addition of glucose in to the reaction environment. Glucose readily permeate the membrane of the polymersome, resulting in the enzymatic conversion of glucose in to lactone and hydrogen peroxide (H_2O_2) as subproduct of reaction. H_2O_2 is subsequently utilized by HRP to convert ABTS to $ABTSC^{\bullet +}$, with maximum efficiency of conversion at pH 5.5 (three times faster when compared to pH 7.5). The one-pot multistep reactions for the three-enzyme system was also monitored using 1, 2, 3, 4-tetra-O-acetyl-b-glucopyranose as substrate. In a first reaction step (the rate-limiting step of reaction), CALB was responsible to hydrolyze acetate groups before the followed reaction pathway by GOx-HRP. The experiments also showed that the conversion reaction in polymersomes was also much more effective than in solution; in the absence of ABTS no enzymatic reaction can proceed. Moreover, the three-enzyme reaction system in polymersomes highlights the clear benefit of encapsulation approach, with a loss of activity of around 87.5% when compared to enzymatic reaction in solution after one month [20].

Cell membranes are formed by natural phospholipids bilayer, to which proteins and other small molecules are bound [21]. Phospholipids can be self-assemble on their own to form vesicles with a spherical phospholipidic bilayer, or liposomes [22]. Liposomes as well as polymersomes are often classified according to their lamellar properties and diameter to form small unilamellar vesicles (SUVs, 20–100 nm), large unilamellar vesicles (LUVs, 100 nm–1 μm), and giant unilamellar vesicles (GUVs, > 1 μm) [23]. Also, the particular hydrophobic lipidic bilayer with a hydrophilic nature of the inner core favors the encapsulation of proteins, enzymes, photosensitizers, vaccines, etc., for a broad range of applications. Although liposomes are good mimics of natural cells membranes they are difficult to handle in comparison to polymersomes basically because of its lower physical stability, higher chemical instability due to the oxidation and hydrolysis of lipidic bilayer membrane [23]. The lower chemical versatility of the membrane is also a limiting factor when practical applications are envisioned. However, liposomes can be also a pathway of synthetic compartmentalization of biomolecules, such as enzymes, and can be prepared according to many protocols of encapsulation [24]. As an example, lipid vesicles are used to obtain living/synthetic hybrid cells to perform a synthetic enzymatic metabolism co-encapsulated in the vesicle [25]. One important feature of lipid vesicles is the permeability of substrates and products of reaction through the membrane. The substantial permeability of the molecules is reported to depend on the dynamics of the lipid membrane and also due to the relative size of uncharged molecules [26]. On the other hand, lipidic bilayer membranes are reported to have selective permeability to charged molecules or other macromolecular species, e.g., generated as products of an enzymatic reaction. The selective permeability of reaction products was reported for the trapping of individual horseradish peroxidase (HRP) biomolecule within LUVs [27], as shown in Fig. 1.

The reaction of the enzyme HRP was studied when encapsulated within surface-tethered lipid vesicles using biotin-avidin chemistry [27]. The LUVs were used to mimic cellular entrapment of HRP (0.5 μmol L^{-1})(Fig. 1a), and the progress of enzymatic reaction was based on the HRP-catalyzed oxidation of the non-fluorescent uncharged molecule Amplex Red (10-acetyl-3,7-dihydroxyphenoxazine) in to the fluorescent product resorufin (7-hydroxy-3H-phenoxazin-3-one) in the presence of H_2O_2. Ample red diffuses freely through the lipid bilayer membrane in to the inner cage of the LUVs, where HRP catalyze its oxidation reaction; H_2O_2 can also cross the bilipidic membrane [28]. However, the product resofurin cannot cross the phospholipids bilayer at neutral pH due to their negative charge causing its accumulation inside the vesicle and was monitored by the increase in the fluorescence signal as the reaction proceed [27]. Figure 1b shows the Microscopic image of a typical sample where the bright spots correspond to the positions of individual vesicles with accumulation of product molecules. The addition of resofurin to the outside of the LUVs did not show the same localized bright spots and no indication of the adsorption by the membrane of the vesicles [27]. Figure 1c–f shows the decline of time traces of fluorescence signals over time of individual vesicles containing enzyme molecules, and not the substrates but the product inhibition is the main cause for the suppression of the reaction. In this case, the Edie-Hofstee plot of v (μmol s^{-1}) as a function of $v/[H_2O_2]$ with a fixed initial concentration of resofurin was used to determine the allosteric inhibition by the product of reaction and a clear prove why enzymatic reaction rates inside vesicles eventually decreases toward close to zero. A possible way to avoid the suppression of enzymatic activity inside vesicles could be the introduction of pores to facilitate the product scape from the inside of the vesicles, preventing its accumulation [27].

Viruses are intracellular parasites that depend on host living cells machineries for their replication [29]. The structure of virus is formed by its protein shell, or capsid, and their genetic material. Virus-like nanoparticles can have different size and shapes according to the chemical environment (e.g., pH and ionic strength), are chemically stable and are similar to the original structure of viruses but without the viral

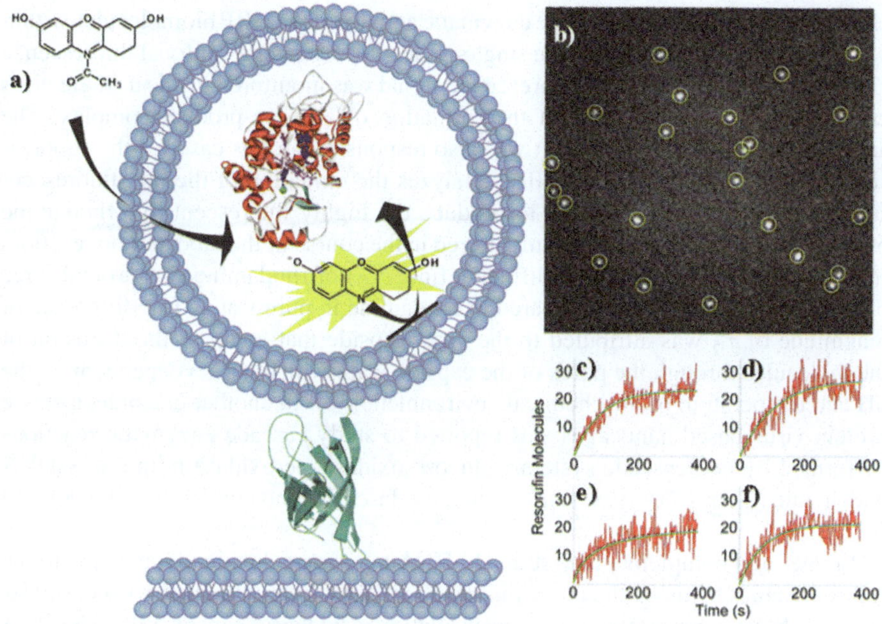

Fig. 1 Individual HRP Trapped in Lipid Vesicles. **a** An individual enzyme molecule within a lipid vesicle. A LUV encapsulating a single HRP molecule is attached to a glass-supported lipid bilayer using biotin-avidin chemistry. Externally added Amplex Red substrate molecules penetrate the vesicle and react with the enzyme, leading to creation of the fluorescent product resorufin that remains trapped in the vesicle interior. Microscope image of a sample with surface-tethered vesicles containing single HRP molecules after incubation with the substrate. The bright spots represent the positions of vesicles where fluorescent resorufin was produced and accumulated. **b–f** Representative time traces of the fluorescence intensity from individual vesicles as a function of time. Following an initial rapid rise, the fluorescence signal approaches a plateau (Reproduced with permission of reference [27])

genome [9]. The basic structure of viruses can be used to insert many kinds of biological molecules [30], such as redox proteins and enzymes, according to non-covalent encapsulation [31], covalent encapsulation [32], charge-mediated encapsulation [33], RNA aptamer-mediated encapsulation [34], scaffolding protein-mediated encapsulation [35], unnatural amino acid-mediated encapsulation [36], and enzyme-mediated encapsulation [37]. Also, the use of virus-like nanoparticles as encapsulating agents provides several advantages for practical applications because they can prevent degradation with the environment and improve the delivery to a specific target [9]. This also resumes the enormous potential of virus-like nanoparticles as drug-delivery systems. In particular, virus-like nanoparticles can also be used to create a favorable condition for encapsulation of enzymes for creation of functional enzymatic virus-like nanoparticles. An important aspect of such capsides is also the porosity of its molecular structure for possible control of reaction product release. As an example, a virus capside was investigated as a nanoreactor for biocatalytic reaction of an individual

HRP enzyme [38]. In this case, the enzymatic activity of the HRP biomolecules within virus capside was examined at the single-molecule level using confocal fluorescence [38]. The relative increase of fluorescence signal was monitored inside a single virus capside over time as a result of the formation of enzyme-product complex. The accumulation of reaction product was also responsible in this case for the suppression of the enzymatic activity. HRP catalyzes the oxidation of the non-fluorescent substrate dihydrorhodamine 6G to produce the highly fluorescent dye rhodamine 6G. The rhodamine 6G was then monitored in the course of the biocatalytic reaction. Interestingly, the characteristic diffusion time (τ_D) of rhodamine was around three orders of magnitude when compared to rhodamine in pure water. The difference of magnitude of τ_D was attributed to the virus capside that hampers the diffusion of the molecules through the pores of the capside wall, which is also dependent on the pH and temperature of the chemical environment [38]. In another example, a single protein virus-based nanocages was reported to study cascade enzymatic reactions performed by two cascade systems, glucose oxidase–peroxidase mimic of ssDNA with hemin (GOx-DNAzyme) and glucose oxidase-glucono kinase (GOx-GCK) [39] (Fig. 2).

Single- and complementary-stranded DNA tags were chemically attached to the redox enzymes by using the cross-linking agent N-[ε-maleimidocaproyloxy] sulfosuccinimide ester, promoting its co-encapsulation inside virus capside [39]. The DNA tags and the outside of the capsides provided tunable system by varying the molecular length of the DNA and buffer salt concentration, with the enzymes non-covalently bound but instead incorporated in to the protein cage of the virus [39]. This protocol was used to create two cascade reaction systems. Figure 2a shows the formation of the capside-like structure with ssDNA (catalytic active in the presence of hemin) and Fig. 2b shows GOx enzyme functionalized with ssDNA. Figures 2c and d show the GCK functionalized with ssDNA and GOx conjugated to GCK, respectively. The enzymatic catalytic activity of both encapsulated cascade systems I (GOx-DNAzyme) and II (GOx-GCK). In the first enzymatic cascade system (Fig. 2e), glucose is converted by GOx enzyme to gluconolactone and H_2O_2 as subproduct of reaction. H_2O_2 is consumed by the DNAzyme in the reaction with ABTS as well as by catalase enzyme. The formation of ABTS cation radical ($ABTS^{\bullet+}$) by the catalytic activity of GOx-DNAzyme was monitored at $\lambda = 410$ nm after the addition of glucose in the reaction system [39]. For the reaction system GOx-GCK, the formation of the end product of reaction nicotinamide adenine dinucleotide phosphate (NADPH) was monitored upon addition of glucose in to the reaction system, at $\lambda = 340$ nm. Figures 2f and g show the Michaelis–Menten kinetics of reaction for cascade systems I and II, respectively, with K_M values similar for both systems. On the other hand, k_{cat} showed a twofold increase for encapsulated systems [39]. Figure 2h shows the kinetic reaction measurements for the production of $ABTS^{\bullet+}$ in the presence of and after pH inactivation of the enzyme catalase. Both reaction cascade reaction pathways also showed that glucose substrate was able to diffuse in to the virus capside shell [39]. DNAzyme exhibit lower catalytic efficiency in the presence of H_2O_2. However, an experimental evidence showed that the subproduct of reaction H_2O_2 can diffuse out of the virus capside and react with enzyme catalase.

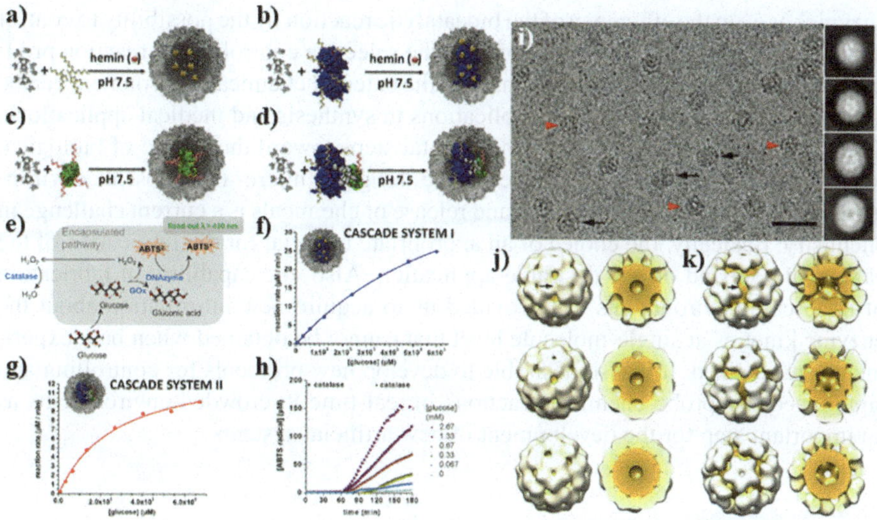

Fig. 2 Enzymatic Cascade Reactions in Virus-based Nanocages. Encapsulation of **a** ssDNA in yellow, **b** GOx, in blue, functionalized with ssDNA, **c** GCK, in green, functionalized with the complementary ssDNA in red, **d** GOx conjugated to GCK. The specific sequence of ssDNA is catalytically active in the presence of hemin. **e** Schematic representation of cascade system I in the presence of a competing enzyme, catalase (encapsulated processes shown in gray boxes). **f** Kinetic measurements of cascade system I; the production of $ABTS^{\bullet+}$ was monitored at $\lambda = 410$ nm at different glucose concentrations. **g** Kinetic measurements of cascade system II; the production of NADPH was monitored at $\lambda = 340$ nm at different glucose concentrations. **h** Kinetic measurements for the production of $ABTS^{\bullet+}$ in the presence of and after pH inactivation of the competing enzyme, catalase. **i** Cryo-electron micrograph of GOx–ssDNA loaded CCMV capsids. Black arrows indicate elongated particles, and red arrowheads indicate irregular particles. Two-dimensional class averages derived from the final 15481 particle data set (inset). Bar, 50 nm. **j** Surface-shaded representation of the outer surface of the class I $T = 1$ capsid (diameter 21.4 nm) viewed along a 2-, 3-, and 5-fold axis of icosahedral symmetry (top to bottom). Models of the class I $T = 1$ capsid, with the front half of the cargo and protein shell removed (right). Protein shell is white, cargo is yellow. Arrow indicates a 2-fold axis of icosahedral symmetry. **k** Surface-shaded representations of the outer surface of the class II $T = 1$ capsid (diameter 22.6 nm) (as in B) (Reproduced with permission of reference [39]. Copyright 2017, American Chemical Society)

2 Final Considerations

It is envisioned that the control of physical and chemical properties of such volume-confined environments can be able to overcome some practical limitations regards to the production of desirable chemicals or even the pathway to which enzymes can catalyze multistep or cascade chemical reactions. In this way, to understand the biocatalytic behavior of enzymes in volume-confined environments is a fundamental step toward practical application in many fields of research. For example, the confinement of biological molecules such as enzymes in micro- or nanoscale reactors strongly influences how the reaction rate of enzymes are carry out, the stability of the enzymes or even the efficiency of the biocatalytic reaction by the possibility to control the local reaction environment. Similarly, the selective control of the reaction products and selective release from the inner to the external chemical environment can be a fundamental step for practical applications in synthesis and medical applications. These are basically some of the fundamental steps toward the mimic of biological reactions in living organisms. However, the design of micro- or nanoscale environments with controllable production and release of chemicals is a current challenge in chemistry. Basically, the choice of an appropriate material for the fabrication of the cage could depend on the desirable application. Also, the capability of fabrication of nanoscale environments can provided us to acquire new information about the enzyme kinetics at single-molecule level that cannot be obtained when bulk experiments are realized. It is also desirable to develop new protocols for controlling and simultaneously probe chemical reactions in real-time in crowded environments as an important step for the development of new artificial systems.

References

1. Chapanian, R., Kwan, D.H., Constantinescu, I., Shaikh, F.A., Rossi, N.A.A., Withers, S.G., Kizhakkedathu, J.N.: Enhancement of biological reactions on cell surfaces via macromolecular crowding. Nat. Commun. **20**(5), 4638 (2014)
2. Goodsell, D.S.: Inside a living cell. Trends Biochem. Sci. **16**(6), 203–206 (1991)
3. Minton, A.P.: How can biochemical reactions within cells differ from those in test tubes? J. Cell Sci. **119**(14), 2863–2869 (2006)
4. Ellis, R.J.: Macromolecular crowding: obvious but underappreciated. Trends in Biochemical Sciences **26**(10), 597–604 (2001)
5. Minton, A.P.: Influence of excluded volume upon macromolecular structure and associations in "crowded" media. Curr. Opin. Biotechnol. **8**(1), 65–69 (1997)
6. Eda, S., Nasibullin, I., Vong, K., Kudo, N., Yoshida, M., Kurbangalieva, A., Tanaka, K.: Biocompatibility and therapeutic potential of glycosylated albumin artificial metalloenzymes. Nat. Catal. **2**, 780–792 (2019)
7. Einfalt, T., Witzigmann, D., Edlinger, C., Sieber, S., Goers, R., Najer, A., Spulber, M., Onaca-Fischer, O., Huwyler, J., Palivan, C.G.: Biomimetic artificial organelles with in vitro and in vivo activity triggered by reduction in microenvironment. Nat. Commun. **9**, 1127 (2018)
8. Tzur-Balter, A., Shatsberg, Z., Beckerman, M., Segal, E., Artzi, N.: Mechanism of erosion of nanostructured porous silicon drug carriers in neoplastic tissues. Nat. Commun. **6**, 6208 (2015)
9. Koyani, R., Pérez-Robles, J., Cadena-Nava, R.D., Vazquez-Duhalt, R.: Biomaterial-based nanoreactors, an alternative for enzyme delivery. Nanotechnol. Rev. **6**(5), 405–419 (2017)
10. LoPresti, C., Lomas, H., Massignani, M., Smart, T., Battaglia, G.: Polymersomes: nature inspired nanometer sized compartments. J. Mater. Chem. **19**(22), 3576–3590 (2009)
11. Che, H., Van Hest, J.C.M.: Stimuli-responsive polymersomes and nanoreactors. J. Mater. Chem. B. **4**(27), 4632–4647 (2016)
12. Egli, S., Nussbaumer, M.G., Balasubramanian, V., Chami, M., Bruns, N., Palivan, C., Meier, W.: Biocompatible functionalization of polymersome surfaces: a new approach to surface immobilization and cell targeting using polymersomes. J. Am. Chem. Soc. **133**(12), 4476–4483 (2011)
13. Ghoroghchian, P.P., Frail, P.R., Susumu, K., Blessington, D., Brannan, A.K., Bates, F.S., Chance, B., Hammer, D.A., Therien, M.J.: Near-infrared-emissive polymersomes: self-assembled soft matter for in vivo optical imaging. Proc. Natl. Acad. Sci. U. S. A. **102**(8), 2922–2927 (2005)

14. Kamat, N.P., Katz, J.S., Hammer, D.A.: Engineering polymersome protocells. J. Phys. Chem. Lett. **2**(13), 1612–1623 (2011)
15. Coustet, M., Irigoyen, J., Garcia, T.A., Murray, R.A., Romero, G., Susana Cortizo, M., Knoll, W., Azzaroni, O., Moya, S.E.: Layer-by-layer assembly of polymersomes and polyelectrolytes on planar surfaces and microsized colloidal particles. J. Colloid Interface Sci. **421**, 132–140 (2014)
16. Hong, Y., Xi, Y., Zhang, J., Wang, D., Zhang, H., Yan, N., He, S., Du, J.: Polymersome-hydrogel composites with combined quick and long-term antibacterial activities. J. Mater. Chem. B **6**(39), 6311–6321 (2018)
17. Martino, C., Kim, S.H., Horsfall, L., Abbaspourrad, A., Rosser, S.J., Cooper, J., Weitz, D.A.: Protein expression, aggregation, and triggered release from polymersomes as artificial cell-like structures. Angew. Chem. Int. Ed. **51**(26), 6416–6420 (2012)
18. Lee, K.Y., Park, S.J., Lee, K.A., Kim, S.H., Kim, H., Meroz, Y., Mahadevan, L., Jung, K.H., Ahn, T.K., Parker, K.K., et al.: Photosynthetic artificial organelles sustain and control ATP-dependent reactions in a protocellular system. Nat. Biotechnol. **36**, 530–535 (2018)
19. Golombek, F., Castiglione, K.: Polymersomes as nanoreactors enabling the application of solvent-sensitive enzymes in different biphasic reaction setups. Biotechnol. J. **15**(11), e1900561 (2020)
20. Vriezema, D.M., Garcia, P.M.L., Sancho Oltra, N., Hatzakis, N.S., Kuiper, S.M., Nolte, R.J.M., Rowan, A.E., Van Hest, J.C.M.: Positional assembly of enzymes in polymersome nanoreactors for cascade reactions. Angew. Chem. Int. Ed. **46**(39), 7378–7382 (2007)
21. Engelman, D.M.: Membranes are more mosaic than fluid. Nature **438**(7068), 578–580 (2005)
22. Szoka, F., Papahadjopoulos, D.: Comparative properties and methods of preparation of lipid vesicles (liposomes). Ann. Rev. Biophys. Bioeng. **9**, 467–508 (1980)
23. Rideau, E., Dimova, R., Schwille, P., Wurm, F.R., Landfester, K.: Liposomes and polymersomes: a comparative review towards cell mimicking. Chem. Soc. Rev. **47**(23), 8572–8610 (2018)
24. Walde, P., Ichikawa, S.: Enzymes inside lipid vesicles: Preparation, reactivity and applications. Biomol. Eng. **18**(4), 143–177 (2001)
25. Elani, Y., Trantidou, T., Wylie, D., Dekker, L., Polizzi, K., Law, R.V., Ces, O.: Constructing vesicle-based artificial cells with embedded living cells as organelle-like modules. Sci. Rep. **8**, 4564 (2018)
26. Chacko, I.A., Ghate, V.M., Dsouza, L., Lewis, S.A.: Lipid vesicles: a versatile drug delivery platform for dermal and transdermal applications. Coll. Surf. B: Biointerfaces **195**, 111262 (2020)
27. Piwonski, H.M., Goomanovsky, M., Bensimon, D., Horovitz, A., Haran, G.: Allosteric inhibition of individual enzyme molecules trapped in lipid vesicles. Proc. Natl. Acad. Sci. U. S. A. **109**(22), 8368–8369 (2012)
28. Bowler, C., Van Montagu, M., Inzé, D.: Superoxide dismutase and stress tolerance. Annu. Rev. Plant Physiol. Plant Mol. Biol. **43**, 83–116 (1992)
29. Azad, K., Banerjee, M., Johnson, J.E.: Enzymes and enzyme activity encoded by nonenveloped viruses. Annu. Rev. Virol. **4**(1), 221–240 (2017)
30. Wilkerson, J.W., Yang, S.O., Funk, P.J., Stanley, S.K., Bundy, B.C.: Nanoreactors: strategies to encapsulate enzyme biocatalysts in virus-like particles. New Biotechnology **44**, 59–63 (2018)
31. Minten, I.J., Hendriks, L.J.A., Nolte, R.J.M., Cornelissen, J.J.L.M.: Controlled encapsulation of multiple proteins in virus capsids. J. Am. Chem. Soc. **131**(49), 17771–17773 (2009)
32. Schoonen, L., Pille, J., Borrmann, A., Nolte, R.J.M., Van Hest, J.C.M.: Sortase A-mediated N-terminal modification of cowpea chlorotic mottle virus for highly efficient cargo loading. Bioconjug. Chem. **26**(12), 2429–2434 (2015)
33. Bundy, B.C., Franciszkowicz, M.J., Swartz, J.R.: Escherichia coli-based cell-free synthesis of virus-like particles. Biotechnol. Bioeng. **100**(1), 28–37 (2008)
34. Fiedler, J.D., Brown, S.D., Lau, J.L., Finn, M.G.: RNA-directed packaging of enzymes within virus-like particles. Angew. Chem. Int. Ed. **49**(50), 9648–9651 (2010)

35. Patterson, D.P., Schwarz, B., Waters, R.S., Gedeon, T., Douglas, T.: Encapsulation of an enzyme cascade within the bacteriophage P22 virus-like particle. ACS Chem. Biol. **9**(2), 359–365 (2014)
36. Shrestha, P., Smith, M.T., Bundy, B.C.: Cell-free unnatural amino acid incorporation with alternative energy systems and linear expression templates. New Biotechnology **31**(1), 28–34 (2014)
37. Schoonen, L., Nolte, R.J.M., Van Hest, J.C.M.: Highly efficient enzyme encapsulation in a protein nanocage: towards enzyme catalysis in a cellular nanocompartment mimic. Nanoscale **8**(30), 14467–14472 (2016)
38. Comellas-Aragonès, M., Engelkamp, H., Claessen, V.I., Sommerdijk, N.A.J.M., Rowan, A.E., Christianen, P.C.M., Maan, J.C., Verduin, B.J.M., Cornelissen, J.J.L.M., Nolte, R.J.M.: A Virus-Based Single-Enzyme Nanoreactor. Nat. Nanotechnol. **2**, 635–639 (2007)
39. Brasch, M., Putri, R.M., De Ruiter, M.V., Luque, D., Koay, M.S.T., Castón, J.R., Cornelissen, J.J.L.M.: Assembling enzymatic cascade pathways inside virus-based nanocages using dual-tasking nucleic acid tags. J. Am. Chem. Soc. **139**(4), 1512–1519 (2017)

Organic Semiconductors as Support Material for Electrochemical Biorecognition: Advantages, Properties, and Biofunctionalization

Nathália Magno Galdino, Lara Fernandes Loguercio, Luiza de Mattos Manica, Carolina Ferreira de Matos, and Jacqueline Ferreira Leite Santos

Abstract Organic semiconductors are among the most promising class of materials for the development of highly efficient bioelectrochemical devices. They have versatility in their structure, morphology, and composition to provide a suitable environment for the immobilization of biomolecules, in addition to booster the electron transfer to the electrode. This chapter summarizes the basic properties of conducting polymers and carbon-based materials (focusing on carbon nanotubes, graphene, and derivatives), as well as synthesis methods to obtain materials with improved properties for application as electrochemical biosensor. Subsequently, it will be discussed strategies for immobilizing the biorecognition elements on the surface of the material.

Keywords Organic semiconductor · Conducting polymer · Carbon nanotube · Graphene · Biofunctionalization · Electrochemical biosensor

1 Introduction

Organic semiconductors have received increasing attention for their application as support material for electrochemical biorecognition due to their versatility in chemical, morphological, and electrochemical behavior, in addition to low cost for production. Tailoring their surface properties and composition during the synthesis allows efficient biomolecules immobilization for an excellent biorecognition performance and simple operation for a variety of analytes. In this type of biosensor, the organic semiconductor may contribute to both sensing and transduction.

N. M. Galdino · L. de Mattos Manica · J. F. L. Santos (✉)
Laboratory of Applied Materials and Interfaces / Institute of Chemistry, Federal University of Rio Grande do Sul, Porto Alegre, RS, Brazil
e-mail: jacqueline.ferreira@ufrgs.br

L. F. Loguercio
Northeast Center for Strategic Technologies, Recife, PE, Brazil

C. F. de Matos
Federal University of Pampa, Caçapava do Sul, RS, Brazil

F. N. Crespilho (ed.), *Advances in Bioelectrochemistry Volume 2*,
https://doi.org/10.1007/978-3-030-95270-9_5

In addition to the chemical composition versatility, these materials can be easily combined with other nanomaterials to improve their biosensor performance. The formation of nanostructured organic semiconductors or the combination with hydrogels, for example, improves the charge transport and mechanical properties, respectively [38, 52, 56]. Furthermore, many strategies have also been reported to improve chemical stability, crystallinity, charge transport, active surface sites, and interactions with biomolecules. For instance, the organic semiconductors can be either functionalized or combined with varied materials (such as metal and metal oxide nanoparticles, molecularly imprinted polymers, and ionic liquids) to obtain nanocomposites with improved chemical stability and large surface area in nano/microporous structures.

The simplicity, high sensitivity, selectivity, robustness, feasibility for mass production, miniaturization, multiplexing, and portability make the organic semiconductor based electrochemical biosensors attractive for the development of point-of-care (PoC) devices [52, 80, 84]. The PoC diagnostics field had been rapidly developed over the past few decades, leading to the commercialization of novel sensing technologies and arrays, consisting of bioreceptors that drive a selective recognition of the target of interest [54, 75]. However, the limited recognition elements stability, the high cost for device fabrication and for obtaining the isolated biomolecules has been encouraged the development of new matrices as support material to allow efficient immobilization of these biomolecules in a suitable environment for biorecognition.

The feasibility and low cost of conducting polymers (CPs) production allow straightforward applications, as they can function as sensing elements and transducers simultaneously. These characteristics enable the simplification of the sensor design, providing a direct electrical readout during the detection of the analytes. The remarkable advantages of CPs over other support materials for electrochemical biorecognition are mainly: (i) improved response to the target, (ii) high sensitivity to small perturbations in environmental conditions, allowing the conversion of a binding event to an electrical readout in a fast measurement, (iii) enhanced sensitivity toward chain conformation alterations, arising from their π structure, (iv) easy processability, (v) low cost, (vi) long lifetime, (vii) the possibility to obtain flexible devices, and finally, (viii) their good electroactivity in buffer solutions and lower/none toxicity effect, turning to be an ideal choice for biosensor fabrication [58, 70] to detect, for example, nucleic acids, proteins, and whole SARS-CoV-2 [58, 70, 74].

Carbon-based materials are a versatile option to compose the chemically modified electrodes for electrochemical biorecognition due to their singular physicochemical characteristics. Furthermore, the electronic properties of some carbon-based materials, such as graphene and carbon nanotubes (CNTs), can improve the biosensor response and allow its miniaturization. The electrode modification with carbon-based materials or nanocomposites containing carbon allotrope might contribute to the sensor response and biomolecules immobilization. The sp^2 carbon-bonded structures provide high electrical conductivity, which enhances the charge transfer on modified electrodes. Additionally, the functionalization of carbon nanomaterials and biocompatibility make them suitable for stabilizing the biorecognition element (preventing unbinding or enzyme denaturation, increasing the biosensor lifetime) and composing wearable devices [20, 26, 69].

Among the carbon-based nanomaterials, graphene and CNTs have been widely applied in the biosensors field for some specific reasons: (i) the possibility to functionalize the surface with any functional group of interest, (ii) the elevated surface area, (iii) high electrical conductivity, and (iv) transducer action converting the interaction between the biorecognition element and the analyte into a measurable signal [26].

Therefore, the outstanding properties of CPs, CNTs, graphene, and derivatives explain the efforts to apply these organic semiconductors as support material for electrochemical biorecognition. Hence, firstly in this chapter, a description of their properties and biofunctionalization are described. The next chapter is focused on describing strategies for the electrochemical biorecognition and finally, it is presented some trends toward the designing of miniaturized high-performance devices, such as by using machine learning and screen-printed electrodes technology.

2 Organic Semiconductors

2.1 Conducting Polymers

The CPs are well known for presenting interesting optical, electrochemical, mechanical, and electrical properties after being submitted to a doping process when charge carriers are formed. The most widely investigated CPs for electrochemical biosensors are polyaniline (PANI), polypyrrole (PPy), and poly (3,4-ethylenedioxythiophene) (PEDOT) [46, 74]. These materials present interesting properties for electrochemical transduction, such as low ionization potential, good electrical conductivity, and high electronic affinity [52].

The semiconductor behavior of these organic materials results from a band gap dependent on their chemical structures. During the polymerization of p-type semiconductors, such as PANI, PPy, and PEDOT, the monomer is oxidized to form a radical cation (polaron) which reacts with neighboring monomers. After successive reactions, the oligomers and eventually the polymers are formed. When a second electron is removed during the polymerization, two situations are expected: the formation of another polaron in another segment of the chain or the removal of an electron from the same segment (after four or five monomeric units in this segment), forming a dication radical (bipolaron) (Fig. 1a).

The charges generated in the polymer backbone are neutralized with counter ions during the doping process (in analogy to inorganic semiconductors, but one should take in mind the difference in dopant concentrations, which is expressively higher for the organic semiconductor) [46, 74]. As a result, distortions in the polymeric chains generate energetic levels between the highest occupied molecular orbital and the lowest unoccupied molecular orbital (Fig. 1b). The formation of polaron/bipolaron in the band gap and the mobility of these charge carriers make these materials conductive [5, 39, 48]. The optical and electronic properties of these aromatic CPs arise from the superposition of p_z orbitals coming from unsaturated bonds.

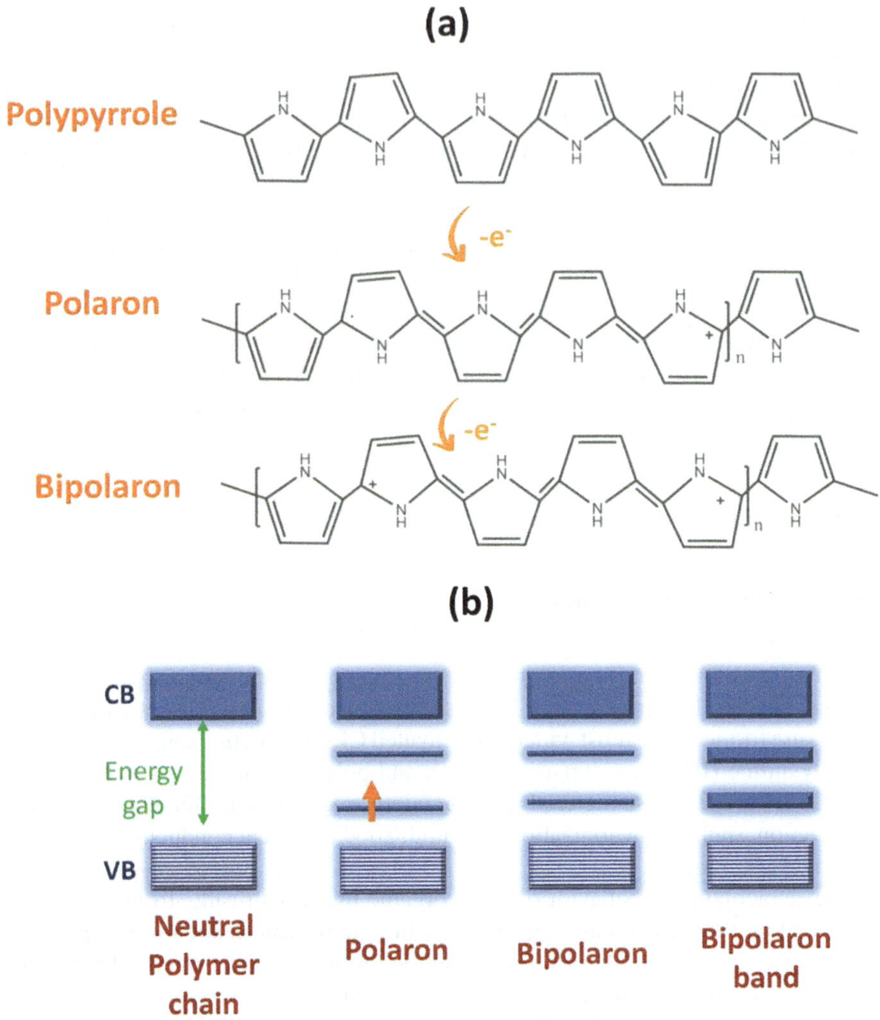

Fig. 1 Formation of **a** polaron and bipolaron in PPy and **b** their energy states

The charge transport in the polymer matrices occurs by intrachain mobility and interchain hopping [10]. The resulting conductivity depends on the number of charges carries and their mobility. Although the CPs usually have a number of charge carriers four to five orders of magnitude higher than in inorganic semiconductors (typically 10^{21} to 10^{23} cm^{-3}), their mobility is lower, limited to only 10^{-4} to 10^{-5} cm^2 V^{-1} s^{-1} (against 10^2 to 10^5 cm^2 V^{-1} s^{-1} in inorganic semiconductors) due to structural defects in the chain [27]. A strategy for improving the conductivity in CPs is controlling the synthesis parameters to obtain ordered materials by chain orientation with dianionic counter ions [23].

Although there are different methods for doping these materials, such as photochemical (leading to high-performance optical materials) and interfacial doping (charge injection without counter ions), the most common methods used to assemble electrochemical biosensors are governed by either chemical or electrochemical oxidation.

In chemical polymerization, oxidizing agents (such as $FeCl_3$) are commonly used, and the produced CP is in powder form. Meanwhile, in electrochemical polymerization, the polymer is deposited onto the anode surface with the application of anodic potentials. The film thickness is controlled during the electropolymerization with the number of cycles and potential range during the cyclic voltammetry (CV) or the time during potentiostat/galvanostatic polarization [46]. The electrochemical synthesis of PPy in powder form has been recently reported [25]. A different approach is a vapor-phase polymerization, where any type of substrate (regardless of its size or shape) can be coated with the polymer. In this case, the morphology is controlled by the temperature and pressure of deposition.

Among all these polymerization methods, the electrochemical stands out with some attractive advantages, including (i) elimination of several steps during the synthesis and purification, (ii) polymerization and film assembling in a few seconds, (iii) any conductive surface may be used for the deposition, such as a transductor and flexible electrodes, and finally, (iv) highly homogeneous films with thickness ranging from few nanometers to micrometers are obtained with a fine control [25, 32, 38, 60, 76].

Some CPs properties are intrinsic and dependent on the polymer identity. For example, PANI conductivity is strongly dependent on pH, the conductivity of emeraldine salt (partial oxidate state) is increased after treatment with acids; on the other hand, PANI presents high electrical resistance when the pH is above 3 (predominating the fully reduced state, leucoemeraldine) [59]. PPy has good conductivity when synthesized in aqueous media, presenting properties strongly dependent on the counter ions [40]. However, the low solubility of thiophene in water usually requires organic solvents such as acetonitrile or propylene carbonate for the synthesis. The use of acetonitrile or propylene carbonate allows large potential windows in electrochemical applications [8]. The dispersion in water may be facilitated by using sulfonated polystyrene as a dopant [72].

Although some CPs properties depend on their identity, the structural, morphological, optical, and electrical properties are easily modulated by choosing suitable polymerization methods, allowing to obtain CPs with desirable properties. The properties can be molded to provide a suitable environment for the biomolecule and to improve the biorecognition kinetics. For example, CPs with a high density of polarons/bipolarons in the surface may allow an efficient immobilization of enzymes in an environment with pH above their isoelectric point (pI). This behavior was observed for the enzyme acetylcholinesterase (AChE), which conformational orientation improves the electron transfer rate from the enzyme to the electrode, as schematically represented in Fig. 2 [42].

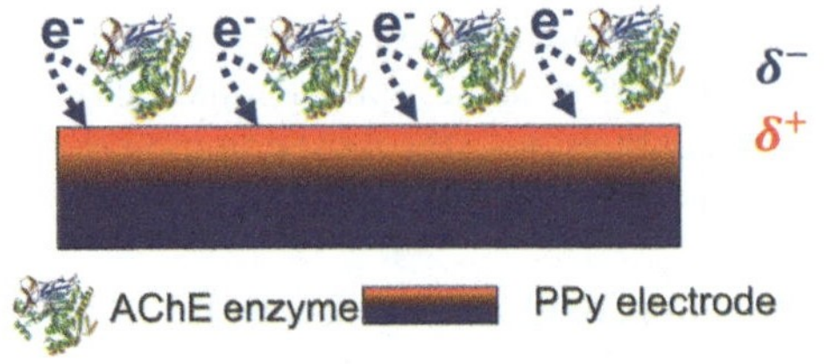

Fig. 2 Schematic representation for the improvement in electron transfer rate between the AChE enzyme and PPy electrode allowed by optimizing the CP surface properties to obtain a suitable environment for the negatively charged enzyme

2.2 Carbon-Based Materials

The carbon-based materials (compounds with mostly carbon–carbon bonds) are usually classified according to the carbon chain organization, which includes 0D (graphene quantum dots, nano-onions, fullerenes, carbon dots, and nanodiamonds), 1D (CNTs, nano-horns, nanofibers), 2D (graphene, graphene oxide, and graphene nanoribbons), and 3D materials (graphite, diamonds, and three-dimensional graphene) [22]. These different structural forms affect the physical–chemical characteristics of these compounds, such as transparency, electrical conductivity, and reactivity. For example, nanodiamonds are electrical isolates but have excellent mechanical and optical properties, high surface area, and tunable surface structures. Moreover, they are suitable for biomedical applications because of their non-toxicity [62]. Meanwhile, graphene and CNTs present excellent electrical conductivity and are widely applied in the bioelectroanalytical field for accelerating the direct electron transfer between the active center of the biomolecule and the underlying electrode. Consequently, the graphene and CNTs based electrochemical biosensors present high sensitivity, selectivity, electron transfer rate, and low detection limit. Therefore, the discussion in this chapter will only focus on these carbon-based materials.

Pristine graphene has one layer of sp^2-hybridized carbon atoms arranged in a hexagonal lattice [55]. It presents the largest theoretical specific surface area among the carbon materials (2630 $m^2\ g^{-1}$), being an ideal support to attach biomolecules. Furthermore, it has high intrinsic mobility, excellent mechanical, optical, and thermal properties. These properties allow, for example, the development of transparent flexible memory, more sensitive, durable, affordable touchscreens, and batteries with better and faster energy storage that can recharge in a few minutes.

Graphene is defined as a zero-band gap semiconductor or semimetal since the density of states is zero at the Fermi energy level [22]. Indeed, the most explored characteristic of this material and its derivatives is the notable electrical behavior. For example, its band gap can be adjusted from 0.26 eV for pristine graphene obtained from epitaxially grown on SiC substrate to zero as the number of layers increases typically to more than four [12]. Additionally, the doping can be tuned electrostatically between n-doping (electrons are the charge carrier in excess) and p-doping (holes are the charge carrier in excess) by the ambipolar electric field effect [22].

The electric behavior and other properties depend highly on graphene crystalline arrangement. Number and size of sheets, along with the amount and type of structural defects on the honeycomb lattice (such as functionalization, edges, doping, C/O ratio) are consequences of the synthetic method and directly influence the band gap energy of graphene. The most frequently used methods to obtain graphene include mechanical or chemical exfoliation of graphite, chemical vapor deposition, epitaxial growth, ultrasonic cleavage, and reduction of graphene oxide (GO) by thermal, electrochemical, chemical, or biological procedures [1]. Due to the huge difficulty of obtaining graphene with high crystallinity (pristine graphene) on a large scale, most studies involving graphene-based electrodes use graphene derivatives such as GO and reduced graphene oxide (rGO).

Schematically, a flat graphene sheet might be stacked into graphite by the van der Waals forces, wrapped up into fullerenes, or rolled into CNTs [21]. As graphene, CNTs have interesting electrical, mechanical, and chemical properties that make them suitable for biosensor applications. Both single-walled carbon nanotubes (SWNTs) and multi-walled carbon nanotubes (MWNTs) are hollow cylinders of sp^2-hybridized carbon atoms with diameters up to 100 nm [22, 87].

The CNTs might behave as metal to moderate semiconductors according to the chirality of the wrap, the nanotubes diameter, and doping [22, 89]. Furthermore, since the curvature introduces a misalignment of the π-orbitals, the reactivity on the convex surface enhances, meanwhile the reactivity in the concave surface decreases after rolling the planar graphene sheet [87]. Besides the electrical properties, other CNTs advantages include excellent electrochemical performance, large surface area, good chemical and thermal stability, easy functionalization, high tensile strength, ultra-lightweight, in addition to favor devices portability and allowing a broad range of working potentials in aqueous solutions [67, 87, 89].

However, one should keep in mind that some defects can be purposefully generated on the CNT and graphene sheet for improving the surface functionalization or modification. The purposeful insertion of defects aims to increase the compatibility of graphene species with other components in nanocomposites. Furthermore, the surface modification of the graphene species can make the material chemically sensitive and soluble, thus allowing its use in detection technology. And, due to its structural characteristics, compared to other carbons, graphene is easier to functionalize in an efficient, reproducible, and homogeneous way. In this sense, several approaches have been used to functionalize graphene and CNTs. Among all, two methods stand out:

i. *Covalent modification* by grafting molecules on the sp^2-carbon atoms of the π-conjugated structure. This functionalization can be achieved via reaction with unsaturated π-bonds of graphene [53] or with p or n heteroatom doping [36]. The direct modification of graphene carbons has the main advantages of forming more stable material, greater control over the functionalization degree, and better reproducibility. However, the covalent functionalization causes losses of the free π electrons associated with sp^2, which constitutes the π-cloud in graphene. This can affect the carrier's mobility, introducing a scattering site

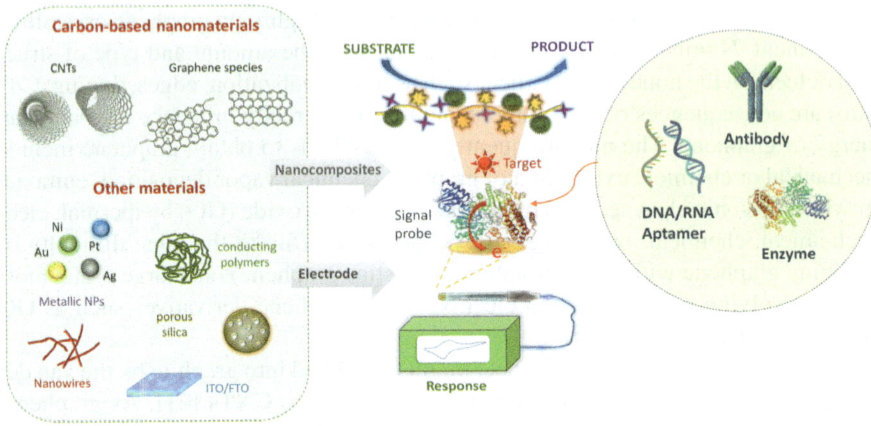

Fig. 3 Schematic representation of the most common nanocomponents in electrochemical biosensors, including the interface of nanostructured electrodes, probes, and biosensing target molecules

and consequently changing the electronic structure and physical properties of graphene [64].

ii. *Non-covalent modification* is based on the adsorption of aromatic compounds or surfactants by π-stacking and hydrophobic interactions with the carbon structure. The main methods for this modification involve the adsorption of surfactant molecules, coating with polymers, and interaction with metallic nanoparticles (NPs) [4] or biomolecules [43]. The advantage of the non-covalent functionalization approach is that by not interfering with the C–C bond of graphene, it preserves the electronic properties of the material, especially its high electrical conductivity. However, as the non-covalent functionalization is non-specific, it lacks a fine control over the degree of surface functionalization, changing the doping density of puddles of electron holes and the presence of scattering sites [82].

In addition to these functionalization methods, the performance of carbon-based materials can be potentially improved when combined with other materials in nanocomposites, for example, when incorporated into CPs matrices. Figure 3 shows a schematic representation for some combinations to assemble the electrochemical biosensors.

2.3 Conducting Polymer and Carbon-Based Materials Composites

Although CPs and carbon-based materials have individually excellent properties as support for biomolecule immobilization and development of electrochemical biosensors, the combination of those materials in composites can result in synergistic effects based on morphological modifications and the electronic interaction between them. As a result, improvements are achieved in the electrical sensing activity, thermal stability, mechanical strength, and chemical properties, resulting in a material with potential for application in different areas, such as solar cells, supercapacitors, transistors, biosensors, and electrochemical sensors [31].

There are different methodologies to obtain these composites. For example, nanocomposites of CPs and carbon materials can be synthesized by (i) electrodeposition of the CP in a pre-prepared CNT or graphene modified electrode; (ii) chemical synthesis; and (iii) in-situ deposition of the CP and the CNTs or reduction of GO in a bare electrode [6]. The interaction mechanisms between the CP and the CNTs are still not fully understood. However, some studies suggest that depending on the mechanism of synthesis, the polymer may covalently functionalize the CNTs [7] or the CNTs may act similarly to anions in the doping process, responsible for maintaining the electrical neutrality in the polymer chain during the doping process [78, 85].

The composites of CP with graphene are used as a nanofiller for the polymer matrix. A new bottom-up synthesis for chemically producing graphene/PANI nanocomposites was reported by Souza [68]. The method is based on chemical reactions at liquid–liquid interfaces, starting from the monomer aniline and benzene, where graphene and the polymer are simultaneously synthesized, forming the nanocomposite with exciting electrical properties [68].

The rGO is also combined with CP in composites. For example, in a composite with PANI, the residual oxygen functional groups in rGO interact with the aniline ions through electrostatic interactions, hydrogen bonding, and π–π interactions. The later interaction can increase the electron transfer, improving the sensing performance of the material [83]. Also, recent studies have shown that GO, even as an electrical insulator, can be used to increase the conductivity of PEDOT:PSS. This increase in conductivity is assigned to changes from PEDOT in benzoid to quinoid structures, promoted by the presence of GO, leading to improvements in the charge transfer rate. Therefore, with a high potential to be applied on the design of electrochemical biosensors [9].

In fact, these CPs-carbon-based materials nanocomposites have been used as electrochemical biosensors due to their unique morphological and electrochemichal properties, providing an excellent environment for the immobilization of several biomolecules, including enzymes, antibodies, microorganisms, DNA, and aptamers. Indeed, there are many articles reporting their use as a matrix for biomolecules immobilization, for instance:

i. with enzymes, e.g.,: for glucose [66, 86], pesticide [11, 17, 29], and vitamin C biosensors [34],
ii. with DNA, e.g.,: for *Mycobacterium tuberculosis* [51, 71], anticancer drug [28], and DNA hybridization biosensors [18, 81],
iii. with aptamers, e.g.,: for detection of dopamine [37, 79], and for pesticide [30],
iv. with antibodies e.g.,: for immunoglobulin G (IgG) [15, 88], vitellogenin (Vtg) [16], aflatoxin B1 [35], and hepatitis B surface antigen (HBsAg) immunosensors [24].

The excellent response provided by carbon materials can be amplified when combined with other nanomaterials, such as NPs, and especially when combined with conducting polymers.

3 Immobilization of the Biorecognition Elements

3.1 Conducting Polymers as Support Material

Advanced support materials with desirable properties for electrochemical biorecognition are obtained with CPs by controlling their surface and bulk properties, allowing the immobilization of a wide variety of biorecognition elements. Besides, improved performance is reached by avoiding fouling surface, not specific binding, keeping the biorecognition sites available, decreasing interferences with physical barriers, or providing an appropriate conformation that enables a fast electron transfer to the electrode during the biorecognition.

The efficient conversion from the biorecognition events to an electrical readout during the transduction strongly depends on the immobilization step, which is expected to be efficient and simple, avoiding steric hindrance to maintain the activity of the recognition probe. There are different suitable methods for the immobilization element, according to its characteristics, the nature of the solid surface, and the transducing mechanism. The reported methods for common recognition elements, such as oligonucleotides, aptamers, antibodies, and enzymes, include electrochemical entrapment, physical adsorption, covalent attachment, and affinity interactions [2, 5].

In the electrochemical entrapment, the biomolecules might be incorporated into CPs during the electropolymerization. This method allows a strong adhesion between biomolecule and polymer film in a single step, in addition, it enables controlling the number of entrapped biomolecules by controlling the film thickness. However, the entrapment may affect the accessibility toward the embedded biomolecules. Furthermore, some CPs require very acidic conditions, high oxidation potential, and counter ions that compete with the biomolecules for the polymer doping sites during the electropolymerization. Thereby, an alternative is controlling the pH of the electrolyte, using biomolecules as counter ions, and applying low potentials during the electropolymerization process [19].

Physical adsorption is based on electrostatic forces between the cationic CPs and anionic biomolecules, hydrophobic and van der Waals forces as well. Therefore, it is essential to evaluate the temperature, pH, solvent type, and net charge of the bioprobe for efficient adsorption. Recently, the literature reported the optimizing pyrrole polymerization to improve the surface conditions for the immobilization of AChE through electrostatic interactions, leading to improved catalytic performance [42].

For covalent binding, N-hydroxysuccinimide/1-ethyl-3-(3-dimethylaminopropyl) carbodiimide (NHS/EDC) are usually used after the polymerization to couple carboxylic acid (–COOH) from CPs to amine ($-NH_2$) groups from the biomolecules. Therefore, any biomolecule with its polypeptide chain can be immobilized onto the CP surface containing carboxylic acid groups.

Affinity attachment is an alternative for covalent immobilization because it also provides strong binding, in addition to reducing the need for chemical reactants. For example, the electrodeposition of biotinylated (antibodies or protein) monomers onto the electrodes, introducing avidin (glycoprotein), builds avidin-biotinylated polymer bridges, or anchoring biotinylated recognition probes onto avidin-biotinylated polymer composites [13, 48, 52, 65, 73].

Efficient immobilization of the biological species is expected with changes in the structure, morphology, and composition of the polymeric matrix, leading to improvements in the biosensor stability, sensitivity, and overall electrochemical performance. CPs micro/nanostructured or nanocomposites can be easily obtained by in-situ or ex-situ methods with several nanomaterials. Nanostructured CPs or porous structures show higher sensitivity and faster response time attributed to the higher active surface area, shorter penetration depth for target molecules and fast diffusion of molecules into the porous structure [19]. Also, the introduction of functional groups, crosslinking near-surface groups, changes in the surface wettability and morphology, in the electrical conductivity, and in the electroactive surface area favor the biomolecule immobilization, allowing the active sites to be closer of the electrode surface, improving the mass or electrons transfer rate [49, 50]. For example, CPs nanocomposite of hydrogels has shown good biocompatibility promoting the immobilization of biomolecules and preserving their bioactivity, excellent processability, which can be easily cast into a thin film with any desired shape as its gelation requires no templates, resulting in a much simple synthetic procedure [63].

A modern strategy to obtain artificial biosensors avoiding the immobilization of biomolecules for biorecognition involves molecularly imprinted polymer (MIP), where the recognition elements can be tailor-made synthetically to match a biomarker target with low cost and simple fabrication [44, 61]. It consists of a template of molecules, functional monomers, and crosslinking reagents. Micro-or nanostructured electrosynthesis of MIP integrated with a small, portable electrochemical device has become a suitable candidate for PoC applications.

For MIP, initially, the polymerization (chemical or electrochemical) occurs around the print species using monomers that are selected for their capacity to form specific and definable interactions with the print species. After, solvent extraction is used to remove entrapped molecules into the polymer. Thus, a microcavity is formed for specifically binding target molecules with the same shape and structure as a lock-and-key, and MIP is ready for use. In addition, in this case, the over-oxidation of CPs is a process that plays a positive role since it creates oxygen-containing groups (mainly carboxyl (–COOH), carbonyl (–CH = O), and hydroxyl (–OH)) very close to the entrapped molecules and these charged groups form a specific environment, which is suitable for the recognition/attachment [14, 61].

Finally, the synergistic effect between CPs and nanomaterials, such as metal oxides, carbon-based materials (CNTs, graphene and their derivatives), metallic NPs, among others, results in matrices with improved properties for biomolecules immobilization with better stability and activity for the bioelement in the solid–liquid interface aiming commercial applications [41, 57, 77] The insertion of these materials in the polymeric matrix can decrease charge transfer resistance (Rct) and mass transfer impedance compared to pristine materials.

3.2 *Carbon-Based Support Materials*

The biofunctionalization of carbon materials is widely applied to obtain sensors with bio-specificity and/or catalytic activity [87]. The carbon nanomaterials are suitable for biofunctionalization due to the high surface-to-volume ratio that allows high biomolecule loading per area. It is exceedingly difficult to characterize exactly how the biorecognition element is anchored on the carbon allotrope matrix because there are many interaction possibilities, some of them occurring simultaneously, as illustrated in Fig. 4. However, molecular dynamic simulations and atomic force microscopy-based single-molecule force spectroscopy are valuable tools for understanding these interactions [33]. Biomolecules might anchor on sp^2 carbon-based materials covalently (surface groups or crosslinkers) or non-covalently (π–π stacking, hydrogen bonds or electrostatic interaction) [33, 87], as detailed below:

i. Covalently with crosslinkers: such as glutaraldehyde (GA), polyethylene glycol (PEG), among others.
ii. Covalently with specific functionalities on carbon frame surface: e.g., amidation reaction between amine groups of proteins or enzymes and carboxylic groups of GO or carboxylic-functionalized CNTs using EDC and NHS.
iii. Electrostatic adsorption: this is the principle for the layer-by-layer assembling, e.g., the graphene has a negative surface charge that might attract enzymes when pH is above the pI.
iv. π–π stacking or hydrophilic interaction between hydrophobic carbon network and aromatic groups may be part of the biomolecule or a linker.
v. Hydrogen-bond: resulting from the interactions with the oxygenated surface groups or others surface functionalities.

The choice for an appropriate methodology for immobilization depends on the characteristics of the carbon-based material, biorecognition element, and environment where the device will be applied. For example, the covalent interactions between carbon materials and biomolecules are more stable, avoiding unbinding; on the other hand, the non-covalent interactions allow to keep the intrinsic properties of the support and preserve the active sites of the biomolecules [33]. Also, the carbon materials can be combined with chitosan, nafion, or other polymers to favor the homogeneity and to prevent leaching of the film. This strategy is widely applied on

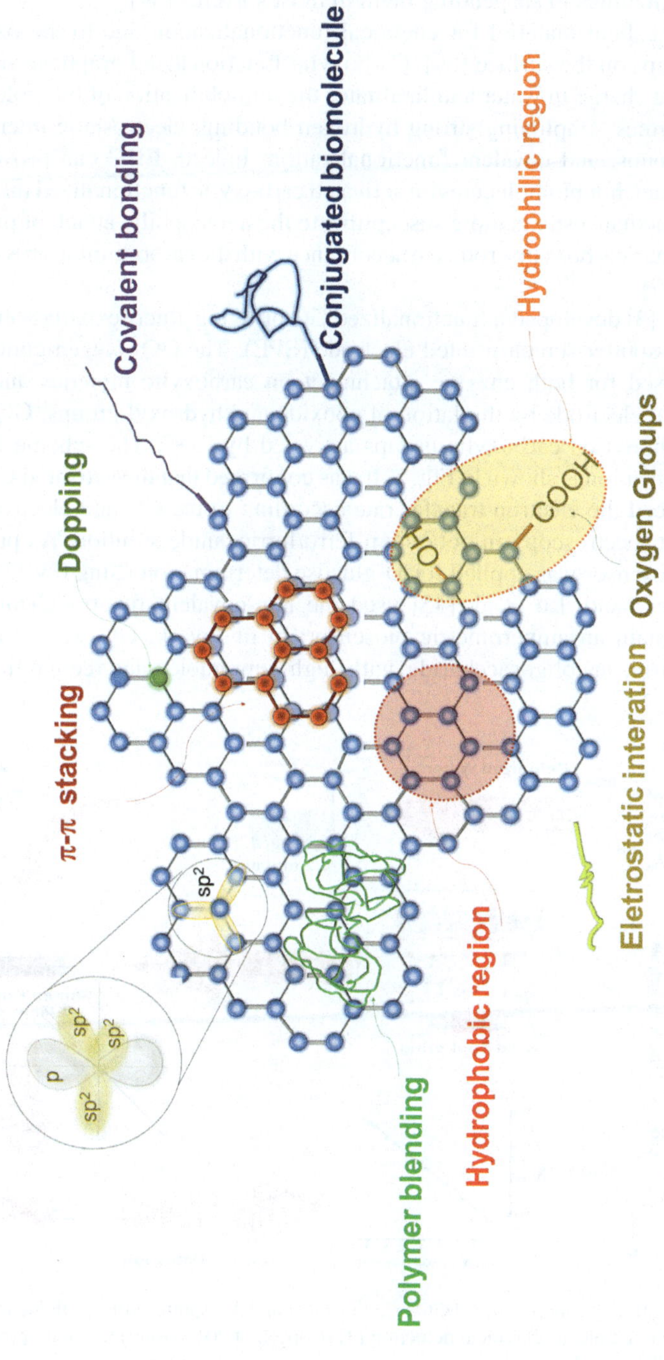

Fig. 4 Schematic illustration of some possibilities for graphene backbone functionalization or modification

electrode modification because both CNTs and graphene have high surface energy what causes difficulties in suspending them in most solvents [89].

GO is an excellent material for chemical functionalization due to the oxygen-containing groups on the surface [69]. Carboxylic-functionalized graphene shows a negative surface charge in water and facilitates the immobilization of biomolecules, especially enzymes, employing strong hydrogen bonding, electrostatic interaction with amino groups, and covalent functionalization. Indeed, EDC can provide an efficient way to anchor biomolecules on activated carboxylic functionalized materials since the intermediate ester is more susceptible to the nucleophilic attack of primary amino groups than carboxylic groups in accordance with the carbodiimide crosslinker chemistry [3, 87].

Akhtar et al. [3] developed a functionalized GO bridging glucose oxidase enzyme (GOx) and Au-sputter screen-printed electrode (SPE). The GO oxygen-containing groups were used for both enzyme attachment on carboxylic moieties and self-assembly on Au electrode by thiolation of epoxide and hydroxyl groups. GOx was covalently anchored on carboxylic groups activated by EDC. The scheme for the electrode modification is shown in Fig. 5. It was confirmed that the activated COOH-GO had enhanced the electron transfer rate according to the CV and electrochemical impedance spectroscopy in potassium ferro/ferricyanide solution. As proof of concept, the biosensor was applied for D-glucose determination using CV [3].

On the other hand, Lu et al. [45] used the non-covalent functionalization of graphene to obtain an amperometric biosensor. In this work, GO was combined with cyclodextrin, an oligosaccharide with high supramolecular recognition and

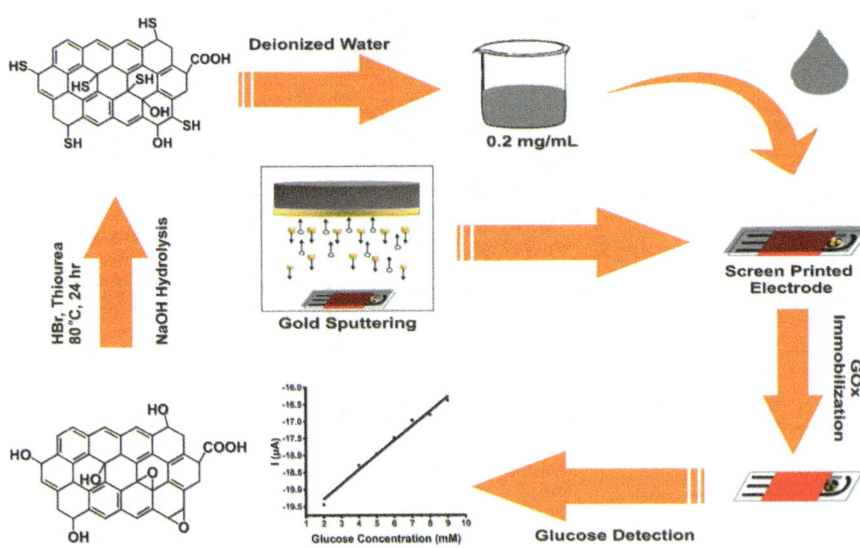

Fig. 5 Representation of functionalized GO bridging GOx and Au-sputter SPE synthetic route and application for glucose electrochemical detection [3] (Copyright 2019 [Reprinted with permission from American Chemical Society])

enrichment capability. Horseradish peroxidase enzyme (HRP) conjugated with 1-adamantane carboxylic acid was self-assembled with cyclodextrin-graphene, presenting good electron-transfer properties and preserving the natural structure of the enzyme (evidenced by the Soret band of HRP in UV–Vis). The amperometric biosensor presents quick response (95% of the steady-state current within 4 s) and long-term stability [45].

In another study, procalcitonin antibodies linked to pyrene-based molecules were attached on highly oriented pyrolytic graphite (HOPG) by π–π staking [47]. According to the authors, the amine groups of the antibody reacted with the activated ester groups of 1-pyrenebutanoic acid succinimidyl ester (PYSE) on HOPG surface. In this case, the HOPG enhances the charge transfer even though the antibody acts as a barrier on the electrode surface. As expected, the antibody-modified electrode showed smaller oxidation and reduction peak currents and higher charge transfer resistance than the HOPG bare sensor. The biosensor was successfully applied to detect the sepsis biomarker procalcitonin by electrochemical impedance spectroscopy, managing to diagnose septic shock [47].

4 Conclusions

Organic semiconductors (specifically conducting polymers, carbon nanotubes, graphene, and derivatives) are interesting materials for the development of high-performance electrochemical biorecognition devices due to their versatility. By controlling the production of the electrodes or synthesis of these materials, one can optimize their interaction with biomolecules, favoring an suitable environment for their conformational stability. In addition, improvements in the biorecognition and electron transfer rate are expected by combining different organic semiconductors or with different nanomaterials.

References

1. Agudosi, E.S., Abdullah, E.C., Numan, A., Mubarak, N.M., Khalid, M., Omar, N.: A review of the graphene synthesis routes and its applications in electrochemical energy storage. Crit. Rev. Solid State Mater. Sci. **45**, 339–377 (2020). https://doi.org/10.1080/10408436.2019.1632793
2. Ahuja, T., Mir, I.H., Kumar D., Rajesh: Biomolecular immobilization on conducting polymers for biosensing applications. Biomaterials **28**, 791–805 (2007). https://doi.org/10.1016/j.biomaterials.2006.09.046
3. Akhtar, M.A., Batool, R., Hayat, A., Han, D., Riaz, S., Khan, S.U., Nasir, M., Nawaz, M.H., Niu, L.: Functionalized graphene oxide bridging between enzyme and Au-sputtered screen-printed interface for glucose detection. ACS Appl. Nano Mater. **2**, 1589–1596 (2019). https://doi.org/10.1021/acsanm.9b00041
4. Arranz-Mascarós, P., Godino-Salido, M.L., López-Garzón, R., García-Gallarín, C., Chamorro-Mena, I., López-Garzón, F.J., Fernández-García, E., Gutiérrez-Valero, M.D.: Non-covalent

functionalization of graphene to tune its band gap and stabilize metal nanoparticles on its surface. ACS Omega **5**, 18849–18861 (2020). https://doi.org/10.1021/acsomega.0c02006
5. Aydemir, N., Malmström, J., Travas-Sejdic, J.: Conducting polymer based electrochemical biosensors. Phys. Chem. Chem. Phys. **18**, 8264–8277 (2016). https://doi.org/10.1039/C5CP06830D
6. Baibarac, M., Gómez-Romero, P.: Nanocomposites based on conducting polymers and carbon nanotubes from fancy materials to functional applications. J. Nanosci. Nanotechnol. **6**, 1–14 (2006). https://doi.org/10.1166/jnn.2006.903
7. Baibarac, M., Baltog, I., Godon, C., Lefrant, S., Chauvet, O.: Covalent functionalization of single-walled carbon nanotubes by aniline electrochemical polymerization. Carbon **42**, 3143–3152 (2004). https://doi.org/10.1016/j.carbon.2004.07.030
8. Bodart, C., Rossetti, N., Hagler, J., Chevreau, P., Chhin, D., Soavi, F., Schougaard, S.B., Amzica, F., Cicoira, F.: Electropolymerized poly(3,4-ethylenedioxythiophene) (PEDOT) coatings for implantable deep-brain-stimulating microelectrodes. ACS Appl. Mater. Interfaces **11**, 17226–17233 (2019). https://doi.org/10.1021/acsami.9b03088
9. Borges, B.G.A.L., Holakoei, S., das Neves, M.F.F., de Menezes, L.C.W., de Matos, C.F., Zarbin, A.J.G., Roman, L.S., Rocco, M.L.M.: Molecular orientation and femtosecond charge transfer dynamics in transparent and conductive electrodes based on graphene oxide and PEDOT:PSS composites. Phys. Chem. Chem. Phys. **21**, 736–743 (2019). https://doi.org/10.1039/C8CP05382K
10. Capaccioli, S., Lucchesi, M., Prevosto, D., Rolla, P.A.: Inter-chain and intra-chain hopping transport in conducting polymers. Phys. Stat. Sol. (c) **1**, 148–151 (2004). https://doi.org/10.1002/pssc.200303611
11. Cesarino, I., Moraes, F.C., Lanza, M.R.V., Machado, S.A.S.: Electrochemical detection of carbamate pesticides in fruit and vegetables with a biosensor based on acetylcholinesterase immobilised on a composite of polyaniline–carbon nanotubes. Food Chem. **135**, 873–879 (2012). https://doi.org/10.1016/j.foodchem.2012.04.147
12. Choi, W., Lahiri, I., Seelaboyina, R., Kang, Y.S.: Synthesis of graphene and its applications: A review. Crit. Rev. Solid State Mater. Sci. **35**, 52–71 (2010). https://doi.org/10.1080/10408430903505036
13. Cosnier, S., Gondran, C.H.: Fabrication of biosensors by attachment of biological macromolecules to electropolymerized conducting films. Analysis **27**, 558–564 (1999). https://doi.org/10.1051/analusis:1999270558
14. Crapnell, R.D., Dempsey-Hibbert, N.C., Peeters, M., Tridente, A., Banks, C.E.: Molecularly imprinted polymer based electrochemical biosensors: overcoming the challenges of detecting vital biomarkers and speeding up diagnosis. Talanta Open **2**, 100018 (2020). https://doi.org/10.1016/j.talo.2020.100018
15. Darain, F., Park, S.-U., Shim, Y.-B.: Disposable amperometric immunosensor system for rabbit IgG using a conducting polymer modified screen-printed electrode. Biosens. Bioelectron. **18**, 773–780 (2003). https://doi.org/10.1016/S0956-5663(03)00004-6
16. Darain, F., Park, S.-U., Chang, S.-C., Shim, Y.-B.: A separation-free amperometric immunosensor for vitellogenin based on screen-printed carbon arrays modified with a conductive polymer. Biosens. Bioelectron. **20**, 1780–1787 (2005). https://doi.org/10.1016/j.bios.2004.07.006
17. Du, D., Ye, X., Cai, J., Liu, J., Zhang, A.: Acetylcholinesterase biosensor design based on carbon nanotube-encapsulated polypyrrole and polyaniline copolymer for amperometric detection of organophosphates. Biosens. Bioelectron. **25**, 2503–2508 (2010). https://doi.org/10.1016/j.bios.2010.04.018
18. Du, M., Yang, T., Li, X., Jiao, K.: Fabrication of DNA/graphene/polyaniline nanocomplex for label-free voltammetric detection of DNA hybridization. Talanta **88**, 439–444 (2012). https://doi.org/10.1016/j.talanta.2011.10.054
19. El-Said, W.A., Abdelshakour, M., Choi, J.-H., Choi, J.-W.: Application of conducting polymer nanostructures to electrochemical biosensors. Molecules **25**, 307 (2020). https://doi.org/10.3390/molecules25020307

20. Galdino, N.M., Brehm, G.S., Bussamara, R., Gonçalves, W.D.G., Abarca, G., Scholten, J.D.: Sputtering deposition of gold nanoparticles onto graphene oxide functionalized with ionic liquids: Biosensor materials for cholesterol detection. J. Mater. Chem. B **5**, 9482–9486 (2017). https://doi.org/10.1039/c7tb02582c
21. Geim, A.K., Novoselov, K.S.: The rise of graphene. Nat. Mater. **6**, 183–191 (2007). https://doi.org/10.1038/nmat1849
22. Georgakilas, V., Perman, J.A., Tucek, J., Zboril, R.: Broad family of carbon nanoallotropes: Classification, chemistry, and applications of fullerenes, carbon dots, nanotubes, graphene, nanodiamonds, and combined superstructures. Chem. Rev. **115**, 4744–4822 (2015). https://doi.org/10.1021/cr500304f
23. Girotto, E.M., Gazotti, W.A., Tormena, C.F., de Paoli, M.-A.: Photoelectronic and transport properties of polypyrrole doped with a dianionic dye. Electrochim. Acta **47**, 1351–1357 (2002). https://doi.org/10.1016/S0013-4686(01)00857-X
24. Hu, Y., Zhao, Z., Wan, Q.: Facile preparation of carbon nanotube-conducting polymer network for sensitive electrochemical immunoassay of Hepatitis B surface antigen in serum. Bioelectrochemistry **81**, 59–64 (2011). https://doi.org/10.1016/j.bioelechem.2011.01.005
25. Istakova, O.I., Konev, D.V., Glazkov, A.T., Medvedeva, T.O., Zolotukhina, E.V., Vorotyntsev, M.A.: Electrochemical synthesis of polypyrrole in powder form. J. Solid State Electrochem. **23**, 251–258 (2019). https://doi.org/10.1007/s10008-018-4129-2
26. Joshi, P., Mishra, R., Narayan, R.J.: Biosensing applications of carbon-based materials. Curr. Opin. in Biomed. Eng. **18**, 100274 (2021). https://doi.org/10.1016/j.cobme.2021.100274
27. Kanatzidis, M.G.: Conductive polymers. Chem. Eng. News **68**, 36–50 (1990). https://doi.org/10.1021/cen-v068n049.p036
28. Karimi-Maleh, H., Tahernejad-Javazmi, F., Atar, N., Yola, M.L., Gupta, V.K., Ensafi, A.A.: A novel DNA biosensor based on a pencil graphite electrode modified with polypyrrole/functionalized multiwalled carbon nanotubes for determination of 6-mercaptopurine anticancer drug. Ind. Eng. Chem. Res. **54**, 3634–3639 (2015). https://doi.org/10.1021/ie504438z
29. Kaur, N., Thakur, H., Prabhakar, N.: Conducting polymer and multi-walled carbon nanotubes nanocomposites based amperometric biosensor for detection of organophosphate. J. Electroanal. Chem. **775**, 121–128 (2016). https://doi.org/10.1016/j.jelechem.2016.05.037
30. Kaur, N., Thakur, H., Prabhakar, N.: Multi walled carbon nanotubes embedded conducting polymer based electrochemical aptasensor for estimation of malathion. Microchem. J. **147**, 393–402 (2019). https://doi.org/10.1016/j.microc.2019.03.042
31. Kaur, G., Kaur, A., Kaur, H.: Review on nanomaterials/conducting polymer based nanocomposites for the development of biosensors and electrochemical sensors. Polym.-Plast. Technol. Mater. **60**, 504–521 (2021). https://doi.org/10.1080/25740881.2020.1844233
32. Lee, J.E., Lee, Y., Ahn, K.-J., Huh, J., Shim, H.W., Sampath, G., Im, W.B., Huh, Y.I., Yoon, H.: Role of co-vapors in vapor deposition polymerization. Sci. Rep. **5**, 8420 (2015). https://doi.org/10.1038/srep08420
33. Li, D., Zhang, W., Yu, X., Wang, Z., Su, Z., Wei, G.: When biomolecules meet graphene: from molecular level interactions to material design and applications. Nanoscale **8**, 19491–19509 (2016). https://doi.org/10.1039/C6NR07249F
34. Li, D., Wen, Y., He, H., Xu, J., Liu, M., Yue, R.: Polypyrrole–multiwalled carbon nanotubes composites as immobilizing matrices of ascorbate oxidase for the facile fabrication of an amperometric vitamin C biosensor. J. Appl. Polym. Sci. **126**, 882–893 (2012). https://doi.org/10.1002/app.36526
35. Linting, Z., Ruiyi, L., Zaijun, L., Qianfang, X., Yinjun, F., Junkang, L.: An immunosensor for ultrasensitive detection of aflatoxin B1 with an enhanced electrochemical performance based on graphene/conducting polymer/gold nanoparticles/the ionic liquid composite film on modified gold electrode with electrodeposition. Sens. Actuators, B Chem. **174**, 359–365 (2012). https://doi.org/10.1016/j.snb.2012.06.051
36. Liu, H., Liu, Y., Zhu, D.: Chemical doping of graphene. J. Mater. Chem. **21**, 3335–3345 (2011). https://doi.org/10.1039/C0JM02922J

37. Liu, S., Xing, X., Yu, J., Lian, W., Li, J., Cui, M., Huang, J.: A novel label-free electrochemical aptasensor based on graphene–polyaniline composite film for dopamine determination. Biosens. Bioelectron. **36**, 186–191 (2012). https://doi.org/10.1016/j.bios.2012.04.011
38. Loguercio, L.F., Alves, C.C., Thesing, A., Ferreira, J.: Enhanced electrochromic properties of a polypyrrole–indigo carmine–gold nanoparticles nanocomposite. Phys. Chem. Chem. Phys. **17**, 1234–1240 (2015). https://doi.org/10.1039/C4CP04262J
39. Loguercio, L.F., Demingos, P., Manica, L.M., Griep, J.B., Santos, M.J.L., Ferreira, J.: Simple One-step method to synthesize polypyrrole-indigo carmine-silver nanocomposite. J. Chem. **2016**, 5284259 (2016). https://doi.org/10.1155/2016/5284259
40. Loguercio, L.F., de Matos, C.F., de Oliveira, M.C., Marin, G., Khan, S., Balzaretti, N.M., Dupont, J., Santos, M.J.L., Santos, J.F.L.: Synergistic interplay of ionic liquid and dodecyl sulphate driving the oxidation state of polypyrrole based electrodes. New J. Chem. **42**, 13828–13835 (2018). https://doi.org/10.1039/c8nj02431f
41. Loguercio, L.F., de Matos, C.F., de Oliveira, M.C., Marin, G., Khan, S., Dupont, J., Teixeira, S.R., Balzaretti, N.M., Santos, J.F.L., Santos, M.J.L.: Polypyrrole/ionic liquid/Au nanoparticle counter-electrodes for dye-sensitized solar cells: Improving charge-transfer resistance at the CE/electrolyte interface. J. Electrochem. Soc. **166**, H3188–H3194 (2019). https://doi.org/10.1149/2.0271905jes
42. Loguercio, L.F., Thesing, A., Demingos, P., de Albuquerque, C.D.L., Rodrigues, R.S.B., Brolo, A.G., Santos, J.F.L.: Efficient acetylcholinesterase immobilization for improved electrochemical performance in polypyrrole nanocomposite-based biosensors for carbaryl pesticide. Sens. Actuators, B Chem. **339**, 129875 (2021). https://doi.org/10.1016/j.snb.2021.129875
43. Lopez, A., Liu, J.: Covalent and noncovalent functionalization of graphene oxide with DNA for smart sensing. Adv. Intell. Syst. **2**, 2000123 (2020). https://doi.org/10.1002/aisy.202000123
44. Lowdon, J.W., Diliën, H., Singla, P., Peeters, M., Cleij, T.J., van Grinsven, B., Eersels, K.: MIPs for commercial application in low-cost sensors and assays—an overview of the current status quo. Sens. Actuators, B Chem. **325**, 128973 (2020). https://doi.org/10.1016/j.snb.2020.128973
45. Lu, L.-M., Qiu, X.-L., Zhang, X.-B., Shen, G.-L., Tan, W., Yu, R.-Q.: Supramolecular assembly of enzyme on functionalized graphene for electrochemical biosensing. Biosens. Bioelectron. **45**, 102–107 (2013). https://doi.org/10.1016/j.bios.2013.01.065
46. Luong, J.H.T., Narayan, T., Solanki, S., Malhotra, B.D.: Recent advances of conducting polymers and their composites for electrochemical biosensing applications. J. Funct. Biomater. **11**, 71 (2020). https://doi.org/10.3390/jfb11040071
47. Mahe, L.S.A., Green, S.J., Winlove, C.P., Jenkins, A.T.A.: Pyrene-wired antibodies on highly oriented pyrolytic graphite as a label-free impedance biosensor for the sepsis biomarker procalcitonin. J. Solid State Electrochem. **18**, 3245–3249 (2014). https://doi.org/10.1007/s10008-014-2588-7
48. Maziz, A., Özgür, E., Bergaud, C., Uzun, L.: Progress in conducting polymers for biointerfacing and biorecognition applications. Sensors Actuators Reports **3**, 100035 (2021). https://doi.org/10.1016/j.snr.2021.100035
49. Meng, L.: Tailoring Conducting Polymer Interface for Sensing and Biosensing. Linköping University Electronic Press, Linköping (2020). https://doi.org/10.3384/diss.diva-169499
50. Meng, L., Turner, A.P.F., Mak, W.C.: Positively-charged hierarchical PEDOT interface with enhanced electrode kinetics for NADH-based biosensors. Biosens. Bioelectron. **120**, 115–121 (2018). https://doi.org/10.1016/j.bios.2018.08.017
51. Mohamad, F.S., Zaid, M.H.M., Abdullah, J., Zawawi, R.M., Lim, H.N., Sulaiman, Y., Rahman, N.A.: Synthesis and characterization of polyaniline/graphene composite nanofiber and its application as an electrochemical DNA biosensor for the detection of *mycobacterium tuberculosis*. Sensors **17**, 2789 (2017). https://doi.org/10.3390/s17122789
52. Moon, J.M., Thapliyal, N., Hussain, K.K., Goyal, R.N., Shim, Y.-B.: Conducting polymer-based electrochemical biosensors for neurotransmitters: a review. Biosens. Bioelectron. **102**, 540–552 (2018). https://doi.org/10.1016/j.bios.2017.11.069
53. Nandanapalli, K.R., Mudusu, D., Lee, S.: Functionalization of graphene layers and advancements in device applications. Carbon **152**, 954–985 (2019). https://doi.org/10.1016/j.carbon.2019.06.081

54. Nayak, S., Blumenfeld, N.R., Laksanasopin, T., Sia, S.K.: Point-of-care diagnostics: Recent developments in a connected age. Anal. Chem. **89**, 102–123 (2017). https://doi.org/10.1021/acs.analchem.6b04630
55. Novoselov, K.S., Geim, A.K., Morozov, S.V., Jiang, D., Zhang, Y., Dubonos, S.V., Grigorieva, I.V., Firsov, A.A.: Electric field effect in atomically thin carbon films. Science **306**, 666–669 (2004). https://doi.org/10.1126/science.1102896
56. Peng, H., Zhang, L., Soeller, C., Travas-Sejdic, J.: Conducting polymers for electrochemical DNA sensing. Biomaterials **30**, 2132–2148 (2009). https://doi.org/10.1016/j.biomaterials.2008.12.065
57. Popov, A., Aukstakojyte, R., Gaidukevic, J., Lisyte, V., Kausaite-Minkstimiene, A., Barkauskas, J., Ramanaviciene, A.: Reduced graphene oxide and polyaniline nanofibers nanocomposite for the development of an amperometric glucose biosensor. Sensors **21**, 948 (2021). https://doi.org/10.3390/s21030948
58. Prakash, S., Chakrabarty, T., Singh, A.K., Shahi, V.K.: Polymer thin films embedded with metal nanoparticles for electrochemical biosensors applications. Biosens. Bioelectron. **41**, 43–53 (2013). https://doi.org/10.1016/j.bios.2012.09.031
59. Rajapakse, R.M.G., Lankeshwara, L.P.P., Careem, M.A.: Factors affecting the electropolymerization of aniline from aqueous solutions. In: International Conference on Science and Technology of Synthetic Metals, Seoul, Korea (South), 24–29 July 1994.
60. Ramanavičius, A., Ramanavičiene, A., Malinauskas, A.: Electrochemical sensors based on conducting polymer-polypyrrole. Electrochim. Acta **51**, 6025–6037 (2006). https://doi.org/10.1016/j.electacta.2005.11.052
61. Ramanavicius, S., Jagminas, A., Ramanavicius, A.: Advances in molecularly imprinted polymers based affinity sensors (review). Polymers **13**, 974 (2021). https://doi.org/10.3390/polym13060974
62. Rehman, A., Houshyar, S., Wang, X.: Nanodiamond-based fibrous composites: A review of fabrication methods, properties, and applications. ACS Appl. Nano Mater. **4**, 2317–2332 (2021). https://doi.org/10.1021/acsanm.1c00470
63. Rong, Q., Han, H., Feng, F., Ma, Z.: Network nanostructured polypyrrole hydrogel/Au composites as enhanced electrochemical biosensing platform. Sci. Rep. **5**, 11440 (2015). https://doi.org/10.1038/srep11440
64. Saini, D.: Synthesis and functionalization of graphene and application in electrochemical biosensing. Nanotechnol. Rev. **5**, 393–416 (2016). https://doi.org/10.1515/ntrev-2015-0059
65. Sassolas, A., Blum, L.J., Leca-Bouvier, B.D.: Immobilization strategies to develop enzymatic biosensors. Biotechnol. Adv. **30**, 489–511 (2012). https://doi.org/10.1016/j.biotechadv.2011.09.003
66. Shrestha, B.K., Ahmad, R., Shrestha, S., Park, C.H., Kim, C.S.: Globular shaped polypyrrole doped well-dispersed functionalized multiwall carbon nanotubes/nafion composite for enzymatic glucose biosensor application. Sci. Rep. **7**, 16191 (2017). https://doi.org/10.1038/s41598-017-16541-9
67. Smart, S.K., Cassady, A.I., Lu, G.Q., Martin, D.J.: The biocompatibility of carbon nanotubes. Carbon **44**, 1034–1047 (2006). https://doi.org/10.1016/j.carbon.2005.10.011
68. Souza, V.H.R., Oliveira, M.M., Zarbin, A.J.G.: Bottom-up synthesis of graphene/polyaniline nanocomposites for flexible and transparent energy storage devices. J. Power Sources **348**, 87–93 (2017). https://doi.org/10.1016/j.jpowsour.2017.02.064
69. Suvarnaphaet, P., Pechprasarn, S.: Graphene-based materials for biosensors: a review. Sensors **17**, 2161 (2017). https://doi.org/10.3390/s17102161
70. Teles, F.R.R., Fonseca, L.P.: Applications of polymers for biomolecule immobilization in electrochemical biosensors. Mater. Sci. Eng. **28**, 1530–1543 (2008). https://doi.org/10.1016/j.msec.2008.04.010
71. Thakur, H., Kaur, N., Sareen, D., Prabhakar, N.: Electrochemical determination of *M. tuberculosis* antigen based on Poly(3,4-ethylenedioxythiophene) and functionalized carbon nanotubes hybrid platform. Talanta **171**, 115–123 (2017). https://doi.org/10.1016/j.talanta.2017.04.063
72. Tomasino, D.V., Wolf, M., Farina, H., Chiarello, G., Feldhoff, A., Ortenzi, M.A., Sabatini, V.: Role of doping agent degree of sulfonation and casting solvent on the electrical conductivity

and morphology of PEDOT:SPAES thin films. Polymers **13**, 658 (2021). https://doi.org/10.3390/polym13040658
73. Torres-Rodriguez, L.M., Roget, A., Billon, M., Bidan, G., Livache, T.: Synthesis of a biotin functionalized pyrrole and its electropolymerization: toward a versatile avidin biosensor. Chem. Commun., 1993–1994 (1998). https://doi.org/10.1039/a804611e
74. Tran, V.V., Tran, N.H.T., Hwang, H.S., Chang, M.: Development strategies of conducting polymer-based electrochemical biosensors for virus biomarkers: Potential for rapid COVID-19 detection. Biosens. Bioelectron. **182**, 113192 (2021). https://doi.org/10.1016/j.bios.2021.113192
75. Vashist, S.K.: Point-of-care diagnostics: recent advances and trends. Biosensors **7**, 62 (2017). https://doi.org/10.3390/bios7040062
76. Wallace, G.G., Smyth, M., Zhao, H.: Conducting electroactive polymer-based biosensors. Trends Anal. Chem. **18**, 245–251 (1999). https://doi.org/10.1016/S0165-9936(98)00113-7
77. Wang, G., Morrin, A., Li, M., Liu, N., Luo, X.: Nanomaterial-doped conducting polymers for electrochemical sensors and biosensors. J. Mater. Chem. B **6**, 4173–4190 (2018). https://doi.org/10.1039/C8TB00817E
78. Wang, J., Musameh, M.: Carbon-nanotubes doped polypyrrole glucose biosensor. Anal. Chim. Acta **539**, 209–213 (2005). https://doi.org/10.1016/j.aca.2005.02.059
79. Wang, W., Wang, W., Davis, J.J., Luo, X.: Ultrasensitive and selective voltammetric aptasensor for dopamine based on a conducting polymer nanocomposite doped with graphene oxide. Microchim. Acta **182**, 1123–1129 (2015). https://doi.org/10.1007/s00604-014-1418-z
80. Wong, C.M., Wong, K.H., Chen, X.D.: Glucose oxidase: natural occurrence, function, properties and industrial applications. Appl. Microbiol. Biotechnol. **78**, 927–938 (2008). https://doi.org/10.1007/s00253-008-1407-4
81. Xu, Y., Ye, X., Yang, L., He, P., Fang, Y.: Impedance DNA biosensor using electropolymerized polypyrrole/multiwalled carbon nanotubes modified electrode. Electroanalysis **18**, 1471–1478 (2006). https://doi.org/10.1002/elan.200603544
82. Yang, G., Bao, D., Liu, H., Zhang, D., Wang, N., Li, H.: Functionalization of graphene and applications of the derivatives. J. Inorg. Organomet. Polym Mater. **27**, 1129–1141 (2017). https://doi.org/10.1007/s10904-017-0597-6
83. Yu, X., Zhang, W., Zhang, P., Su, Z.: Fabrication technologies and sensing applications of graphene-based composite films: advances and challenges. Biosens. Bioelectron. **82**, 72–84 (2017). https://doi.org/10.1016/j.bios.2016.01.081
84. Zan, X., Bai, H., Wang, C., Zhao, F., Duan, H.: Graphene paper decorated with a 2D array of dendritic platinum nanoparticles for ultrasensitive electrochemical detection of dopamine secreted by live cells. Chem. Eur. J. **22**, 5204–5210 (2016). https://doi.org/10.1002/chem.201504454
85. Zengin, H., Zhou, W., Jin, J., Czerw, R., Smith, D.W., Jr., Echegoyen, L., Carrol, D.L., Foulger, S.H., Ballato, J.: Carbon nanotube doped polyaniline. Adv. Mater. **14**, 1480–1483 (2002). https://doi.org/10.1002/1521-4095(20021016)14:20%3c1480:AID-ADMA1480%3e3.0.CO;2-O
86. Zhong, H., Yuan, R., Chai, Y., Li, W., Zhong, X., Zhang, Y.: In situ chemo-synthesized multi-wall carbon nanotube-conductive polyaniline nanocomposites: characterization and application for a glucose amperometric biosensor. Talanta **85**, 104–111 (2011). https://doi.org/10.1016/j.talanta.2011.03.040
87. Zhou, Y., Fang, Y., Ramasamy, R.P.: Non-covalent functionalization of carbon nanotubes for electrochemical biosensor development. Sensors **19**, 392 (2019). https://doi.org/10.3390/s19020392
88. Zhu, Y., Koh, W.C.A., Shim, Y.-B.: An Amperometric immunosensor for IgG based on conducting polymer and carbon nanotube-linked hydrazine label. Electroanalysis **22**, 2908–2914 (2010). https://doi.org/10.1002/elan.201000394
89. Zhu, Z.: An overview of carbon nanotubes and graphene for biosensing applications. Nano-Micro Lett. **9**, 25 (2017). https://doi.org/10.1007/s40820-017-0128-6

Conducting Polymers and Carbon-Based Materials in Biosensor Applications

Fabio Ruiz Simões, Gabriela Martins de Araújo, and Milton Alexandre Cardoso

Abstract Conducting polymers (CPs) and carbon-based materials are examples of materials used in modern biosensors that enable scientific advances in different areas of concentration. This chapter discusses the main characteristics of the CPs such as polyaniline (PAni), polypyrrole (PPy), and polythiophene and carbon-based materials such as fullerenes, carbon nanotubes, and graphene besides reporting the recent advances for biosensor applications. CPs have attracted much interest, especially in biosensors due to the specificity that can be reached with the immobilization of biomolecules its structures such as enzymes. Already the sensitivity of the biosensors can be increased by producing the CPs as nanofibers and nanostructured films. Carbon-based materials such as carbon nanotubes have also been widely used in biosensor applications which act as highly efficient transducers supporting enzymes. Graphene has been used in the development of biosensors in different areas due to its high specific area, high conductivity, and low thickness. These materials have been explored in the development of biosensors used in clinical, environmental, and food applications. Biosensors based on CPs and carbon-based materials are low-cost, simple, specific, sensitive as well as portable devices and with miniaturized size enabling scientific advances in bioelectrochemistry.

Keywords Conducting polymers · Carbon-based materials · Biosensors · Bioelectrochemistry

1 Introduction

Biosensors are electronic devices that respond to biochemical stimulus and generate an electrical impulse that can be used to quickly detect chemical or biological species.

F. Ruiz Simões (✉)
Institute of Marine Sciences, Federal University of São Paulo (UNIFESP), Santos, São Paulo, Brazil
e-mail: fabio.simoes@unifesp.br

G. Martins de Araújo · M. A. Cardoso
Graduate Program in Chemistry: Sustainability Science and Technology, Federal University of São Paulo (UNIFESP), Diadema, São Paulo, Brazil

F. N. Crespilho (ed.), *Advances in Bioelectrochemistry Volume 2*,
https://doi.org/10.1007/978-3-030-95270-9_6

They are currently used in many applications, such as in the areas of health [1–3], food [4–6], and the environment [7, 8]. In general, they are made up of an electronic part, a transducer, and a sensor element that interacts with the analyte. The sensor element can be biological material, such as cells, organelles, proteins, antibodies, among others, which is immobilized on the sensor surface [9].

Biosensors can be thermometric [10], optical [11], piezoelectric [12], or electrochemical [1, 7, 13–15]. Among them, the electrochemical biosensors stand out, which act through the diffusion of electroactive species or in ionic charge. Electrochemical biosensors are the most popular and are used extensively to monitor blood glucose in blood samples for diabetes control [16]. Some of the reasons that favor its application are the good selectivity and sensitivity, agility in carrying out the analysis, usually do not require prior sample treatments, and have low consumption of reagents, which make it more sustainable. Furthermore, they can be amperometric, measuring variations in electrical current due to redox reactions; conductometric, measuring the conductance in catalytic reactions; or potentiometric, measuring the potential difference between electrodes with constant current [17]. Currently, amperometric biosensors are the most used, as the concentration of the analyte is proportional to the current intensity. Thus, it is common to use differential pulse voltammetry (DPV) to quantify this variation during the analysis of the electrochemical amperometric biosensor against the analyte.

Biosensors have been known since 1962, for the work of Clark and Lyons [18]. And they have been increasingly studied in association with modern materials. Conducting polymers (CPs) are examples of materials used in modern biosensors because they respond reversibly between conductive and insulating states [19]. Through oxidation and reduction reactions, the electrochemical biosensor can identify and quantify the analyte [9]. Other modern materials that have risen with technological development are nanotechnological materials. There are many materials whose nanoparticles are applied in sensory devices. However, carbon allotropes are noteworthy because they are low-cost materials (except diamond) that vary in their conductive state according to their carbon hybridization [20]. Carbon nanotubes and graphene are carbon allotropes widely used in the manufacture of sensors and biosensors, but they are not necessarily used alone.

2 Conducting Polymers

Conducting polymers (CPs) are electrically conductive materials composed of organic polymers. The association of the electrical conductivity of metals with the mechanical properties of polymers has been studied, since the mid-1950s, with the incorporation of conductive loads inserted into the polymer matrices. These polymers were called "extrinsically conducting polymers". From the 1970s onwards, a new class, the "intrinsically conducting polymers", has been developed, where the electrical conductivity is independent of the incorporation of conductive loads and its properties are similar to inorganic metals and semiconductors [21]. Intrinsically, CPs represent a class of "synthetic metals" that combine the chemical and mechanical properties of polymers with the electronic properties of metals and semiconductors [22].

2.1 *Main Characteristics of Intrinsically Conducting Polymers*

The first "intrinsically CPs" was obtained in 1977 by H. Shirakawa (University of Tsukuba) with the collaboration of the professors A. MacDiarmid (University of Pennsylvania) and A. Heeger (University of California, Santa Barbara), by exposing polyacetylene in its insulating form (conductivity, $\sigma = 10^{-5}$ S cm^{-1}to doping, oxidizing or reducing agents, making it an intrinsic electrical conductor ($\sigma = 10^{-2}$ S cm^{-1}) [19]. Due to the success of polyacetylene synthesis, the research field of "CPs" had started. In 2000, H. Shirakawa, A. MacDiarmid, and A. Heeger were awarded with the Nobel Prize in Chemistry.

CPs are also known as "conjugated polymers", as they have conjugated double bonds in their structure (single bonds "σ" alternated with double bonds "σ e π") with sp^2 hybridization between the carbon bonds, as shown in Fig. 1 [19]. However, for a polymer to be considered an electrical conductor, it is not enough to have conjugated double bonds. It is also necessary to cause some disturbance, by either the removal or the insertion of electronic charges (oxidation or reduction). This process is called doping [23].

The electrical properties of CPs can be reversibly changed on a wide range, from insulator to metallic conductor. Conductivity can be increased several times by "doping", using substituents that can be oxidizing or reducing agents, or radical's electron donors or acceptors. This "doping process" can be performed by chemical methods of direct exposure of the polymer to the charge transfer agents (doping) in the gaseous

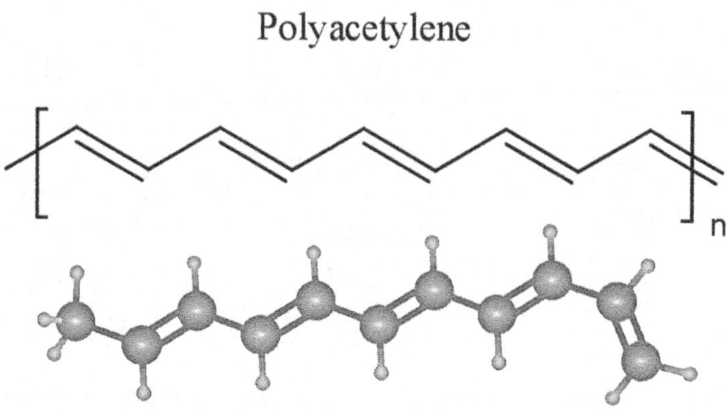

Fig. 1 Polyacetylene molecular structure

phase or solution, or by electrochemical oxidation or reduction. The term "doping" is used in analogy to crystalline inorganic semiconductors, since in both cases doping is random and does not affect the material structure [23–25].

Doping agents can be neutral molecules, acids, salts, bases, organic or inorganic compounds that can easily ionizable. The doping process plays an important role in the stability and conductivity of the CPs, since it leads to the formation of defects and deformations in the polymer chain known as polarons and bi-polarons, which are responsible for the displacement of charges and for the conductivity increasing [26, 27]. The electrons π delocalized along the polymeric chain give conductive properties and provide the ability to support the charge carriers, with high electrical mobility along with the polymer structure [27].

The advantages of CPs are not limited to electronic and/or optical properties of metal and inorganic semiconductors, but also their flexibility, processability, and mechanical properties [21]. Thus, since their discovery, conductive polymers have attracted great attention in several applications such as charge dissipating materials, optoelectronics, solar cells, supercapacitors, batteries, sensors, and biosensors [21]. The most common intrinsically CPs are polyaniline (PAni), polypyrrole (PPy), polythiophene, polyacetylene, poly(p-phenylene), and poly(p-phenylene sulfonated), Fig. 2. The molecular structure of these conjugated polymers exhibits

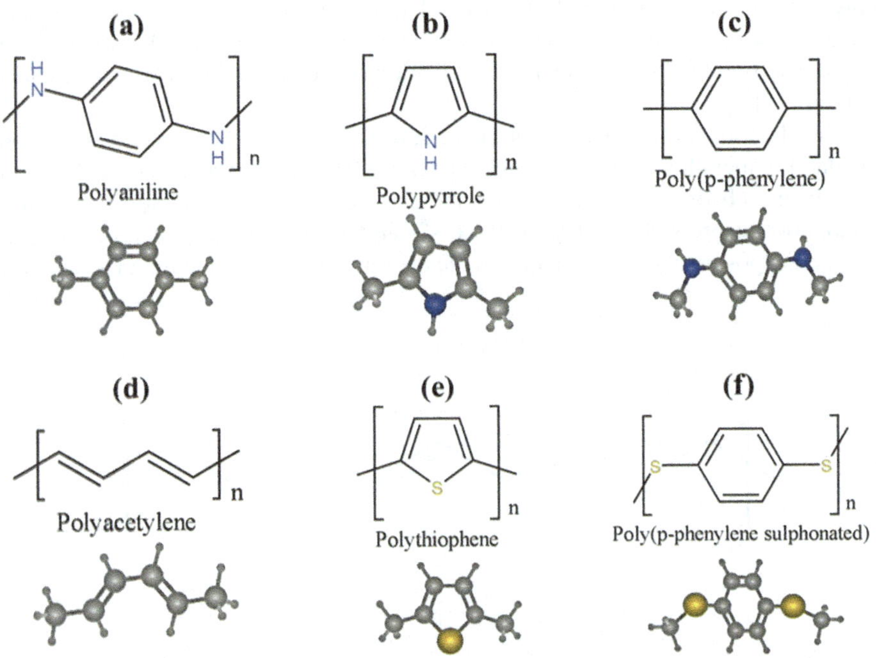

Fig. 2 Molecular structures of common conducting polymers. **a** Polyaniline; **b** polypyrrole; **c** poly(p-phenylene); **d** polyacetylene; **e** polythiophene; and **f** poly(p-phenylene sulfonated)

properties such as low ionization potential and high electronic affinity, and as a result, they can be easily reduced or oxidized [28].

Due to its characteristics such as the electrical conductivity that can be modulated on a wide range, the CPs have been extensively used in the development of electrodes aiming applications such as sensors [29], biosensors [30, 31], batteries [32], and supercapacitors [33]. In sensor and biosensor applications, the CPs are used as active transducer materials [29–31]. However, in sensor applications, the CPs are sensitive but not specific [29]. The sensitivity can be increased by producing as nanofibers [34, 35] and nanostructured films [36]. The specificity can be reached with the addition of functional groups or with immobilized molecules in its structures such as enzymes [37, 38], porphyrins [39], among others.

CPs can act as excellent materials for the immobilization of biomolecules and rapid electron transfer for the fabrication of efficient biosensors [41]. At the end of the twentieth century, the first experimental results that join biosensors and CPs were released. The CPs most used in biosensors are PAni and PPy.

2.2 *Applications of Intrinsically Conducting Polymers at Biosensors*

PAni is an intrinsically CP and has great potential for biomedical applications [41]. It is one of the well-known CPs because it has high conductivity, good environmental stability, and low toxicity and is hydrophilic, having high biocompatibility. Furthermore, PAni can be easily synthesized in aqueous solutions, having a low cost. Its different forms of oxidation allow for variations in stability, conductivity, and color, factors that arouse the interest of several industrial fields [42]. The electrical conductivity of PAni can be regulated, as it is related to its degree of protonation, type of dopant, redox state, and temperature. It has a reversible doping/undoping capability and features two redox pairs, facilitating charge transfer between enzyme and polymer in the biosensor. Thus, PAni acts as a mediator of electron transfer in biosensors, dispensing with the use of other diffusional mediators [41].

Hamid et al. (2018) [43] developed an electrochemical biosensor for the detection of danazol (Dz) which is a drug used in the treatment of endometriosis. The authors produced non-enzymatic biosensors based on PAni electropolymerized onto carbon electrodes. After characterizing the films, they studied the behavior of the electropolymerized film using Cyclic Voltammetry (CV). They used 0.88 M of aniline and hydrochloric acid (HCl) as a dopant, varying the doping concentrations in 1.0, 1.2, and 1.5 M. Their results showed that the oxidation and reduction peaks grow with the increased with successive scans of potential indicating the growth of the PAni polymer chain. The current intensity signal was greater with doping performed with 1 M HCl reaching 44.67 mA/cm^2 at a potential of 0.2 V. By dissolving Dz in methanol (0.01 M), a stock solution was prepared, later diluted to various concentrations, and stored in the dark and refrigerated. These solutions were used as supporting electrolytes for electrochemical analysis of DPV, at a potential from -0.4 to 0.4 V. The assay parameters were optimized and fixed. A study was also carried out with electrochemical impedance spectroscopy (EIS) in phosphate buffer (PBS). EIS showed that the polarization resistance (R_p) also called the charge transfer resistance (R_{ct}) increases with increasing Dz concentrations. The use of this technique is interesting as it explores the use of EIS for analyte analysis and not just for film characterization. DPV performed using PBS (pH = 1) as supporting electrolyte showed that the highest current intensity obtained was 1.671 mA for a concentration of 2 nM.

Asmatulu et al. (2019) [44] developed nanobiosensors based on PAni. The authors fabricated porous PAni nanofibers in 4 diameters through the electrospinning process. The nanobiosensor was manufactured onto a printed circuit platform with an auxiliary electrode and interdigitated working electrode, deposited onto a silicon layer with electrospinning PAni nanofibers. These nanofibers were deposited on top of the gold (Au) interdigitated electrodes, developing four detection platforms. Polydimethylsiloxane manifold (PDMS) was used to mitigate contamination risks and prevent evaporation, so PAni porous nanofibers were integrated into Au interdigitated microelectrodes forming a lab-on-chip device. After treating the platform with ethanol, cleaning with ultrapure deionized water (DI), and drying with nitrogen gas, the biosensor collector was injected with PBS solution, followed by EIS and functionalization of the Au surface. The optimized concentration of cyclooxygenase-2 (COX-2) antibody was injected into the collector, which was incubated. Bovine serum albumin (BSA 200 μL) has been added, aiming to prevent the adsorption of non-specific antigen. EIS performed after insertion of BSA represents the interaction between the antigen and the antibody. After, a COX-2 solution (200μL) was prepared and injected into the collector, which acted as a blocking peptide. EIS measurements were performed alternated by washing the device with PBS and then injecting a higher concentration of COX-2 solution, varying the concentrations of 10 fg mL^{-1} to 1.0 μgmL^{-1}. The results showed that the developed nanobiosensors detected COX-2 at concentrations as low as 0.01 $pgmL^{-1}$ either in the buffer solution or in the human serum samples. Through EIS, the authors proved the selectivity of the device and showed that the PAni nanofiber with the smallest mean diameter (256 nm) showed greater sensitivity.

The potentiometric flexible tattoo biosensor manufactured by Mishra et al. (2018) [45] uses the pH sensitivity properties of PAni to monitor the release of protons from the diisopropyl fluorophosphate (DFP) analyte during hydrolysis by the enzyme organophosphate hydrolase (OPH). The sensor was printed in a skull shape. One eye corresponds to the working electrode printed with carbon; the other corresponds to the reference electrode printed with Ag/AgCl. After the preparation of the printed tattoo sensor, PAni was electropolymerized on the working electrode and a Nafion/OPH layer and PVA hydrogel were deposited by drop-casting. DFP is a toxic organophosphate (OP) pesticide; therefore, the necessary precautions were taken when carrying out the experiment in liquid and vapor phases. Initially, a stock solution of DFP (0.1 M) was prepared. When injected, the enzymatic hydrolysis of DFP occurs when it reaches the outer layer of the Nafion/OPH biosensor and the DFP is absorbed and diffused. There is a release of ions that diffuse toward the electropolymerized layer of PAni, detecting the pH and protonating the active sites. Variations in the obtained potentials generate the quantification of DFP. Fatigue tests were also performed on the biosensor that did not show structural damage. There was linearity in the pH ranges and stability in the response. The biosensor presents a fast response, with efficient analytical performance and good reproducibility, and can be applied to detect OPs pesticides in direct contact with the skin. This type of flexible sensor, with good mechanical strength and good analytical response, expands the application possibilities of electrochemical biosensors.

PPy is also widely used as an electrochemical sensor and there is great interest in its use in biomedical areas, as PPy has high biocompatibility and good electrical conductivity. Thus, several types of biosensors use this CP, such as immunosensors [46] and enzymatic sensors [47]. PPy has the advantage of enlarging the surface area of the electrode and, at the same time, keeping the interfacial resistance low. These factors contribute to the increased sensitivity of the resulting device, which is expressed in the increased charge density of the antibody in a biosensor, for example [48].

Among some works that use it is worth mentioning the work of Ayenimo and Adeloju (2017) [49] who developed an amperometric glucose biosensor based on PPy. Glucose sensors are very important. They allow monitoring of glucose, delaying or controlling diabetes and complications related to it. In this work, platinum disk electrodes (2.0 mm^2) were used, pre-treated by CV in an H_2SO_4 solution (1.0 M) until obtaining a constant electrochemical response. The film was prepared using a 0.2 M of pyrrole monomer and 300 U $mL^{-1}$of immobilized glucose oxidase (GOx), prepared galvanostatically by electropolymerization. An outer layer based on Py (0.1 M) and KCl (0.05 M) was also made, forming the outer ultrafine layer (PPy-Cl 7.9 nm) also prepared galvanostatically. The biosensor was tested on fruit juices diluted in DI by standard addition. Thus, a bilayer biosensor was developed containing a film of PPy with GOx and another external film of permselective PPy-Cl. The electrochemical behavior of the PPy-GOx biosensor without and with the inclusion of the outer layer was studied with and without glucose. By adding the second layer of PPy-Cl, the first was not compromised, obtaining well-defined and similar redox peaks. Furthermore, the second layer contributed to the retention of GOx and is effective in eliminating the influence of uric acid, glycine, ascorbic acid, and glutamic acid interferences. Thus, the inclusion of the second film made was possible to minimize interference in the amperometric detection of glucose. By raising the glucose concentration, the results exhibited increased linearity, exhibiting superior selectivity for glucose detection. Recovery studies reached levels of 90–101% recovery. The results showed that the biosensor achieved good reproducibility and stability with a low detection limit (26.9 μM) and good sensitivity. In addition, the study applied the bilayer biosensor to detect glucose in various fruit juices and was successful.

Akdag et al. (2020) [50] developed a PPy biosensor with chitosan (Chi) and acetylcholinesterase (AChE) to determine acetylthiocholine (ATCh) and the pesticide paraoxon. The PPy was initially electropolymerized on a platinum (Pt) electrode and subsequently covered with a Chi solution for enzyme immobilization. After neutralizing the electrode, it was incubated with a glutaraldehyde solution that was deposited on the electrode surface for 2 h. The biosensor was characterized and its analyte detection performance evaluated by DPV. The results proved the electropolymerization of PPy on a Pt electrode. The detection limit obtained was 0.45 μM for ATCh and 0.17 nM for paraoxon, which shows that the manufactured biosensor was more efficient in determining the pesticide. However, the authors were able to immobilize AChE on the electrode surface and detect both analytes.

3 Carbon-Based Materials

Due to advances in nanoscience and nanotechnology, the miniaturization of systems and devices has many advantages such as less energy consumption, lower cost, and improved functionalities compared to other materials. Carbon-based materials can have different allotropic forms with different hybridizations (sp^3, sp^2, and sp). Thus, the development of nanoscience and nanotechnology has been deeply attached to the development of new carbon-based materials, and among the allotropes, we can highlight those based on graphite structures such as fullerene, carbon nanotubes, and graphene (Fig. 3) [20].

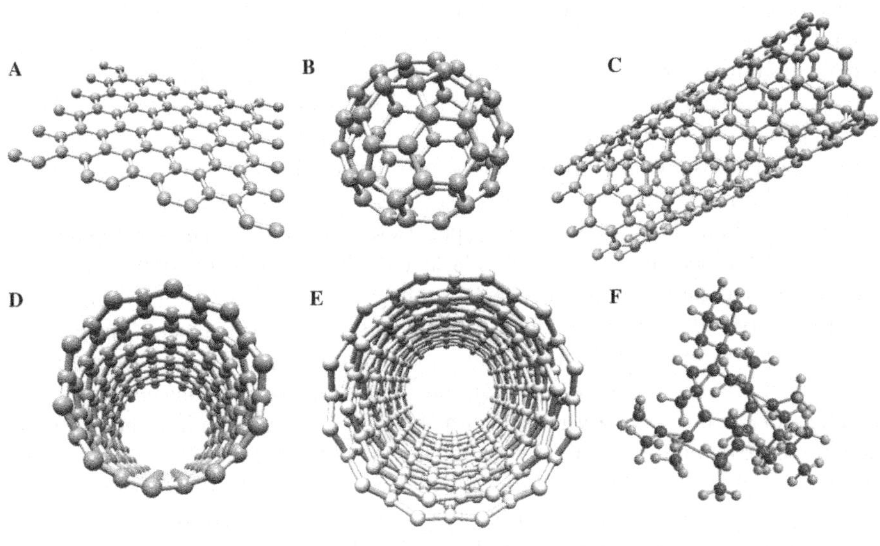

Fig. 3 Carbon-based materials: **a** graphene; **b** fullerene; **c** SWCNT side view; **d** SWCNT apical view; **e** MWCNT; **f** diamond crystal structure [51]

3.1 Main Characteristics

Fullerenes were discovered by Smalley (Rice University) and its collaborators, Kroto and Curl in the mid-1980s. The molecules are structured in the "cages" forms built up of sp^2 carbon hybridizations forming a network of pentagons and hexagons. Their discovery allowed new hypotheses about carbon nanostructures and how the architectures formed from sp^2 carbon units, even with simple geometries, can originate new structures with distinct properties [20, 52, 53].

The carbon nanotubes (CNTs) were discovered by Iijima in 1991 [54] and are materials consisting of carbon cylinders with nanometric diameters and special properties such as high electrical conductivity, excellent chemical stability, and high values of mechanical strength modulus and specific area. There are two distinct types: single-walled carbon nanotubes (SWCNT) formed a single layer and multi-walled carbon nanotubes (MWCNT) with more than one layer of graphite sheet (concentric tubes) (Fig. 3d and e). The main technique used in the production of carbon nanotubes is "Chemical Vapour Decomposition" consisting in the catalytic decomposition of ethylene or acetylene at high temperatures in the presence of a metallic catalyst [55, 56].

Due to its characteristics such as high specific area and charge storage capacity, the carbon nanotubes have been used in the development of nanoelectronics devices [57], batteries [58], hydrogen storage [59], among others. Also, due to their ability to improve electron transfer reactions, contributing to the improvement of reaction kinetics and being able to catalyze oxidizing electrochemical reactions, carbon nanotubes have also been widely used in electrochemical sensors [60] as well as act as highly efficient transducers supporting enzyme and other molecules in biosensor applications [61–63].

Graphene is the first two-dimensional (2D) atomic crystal (hexagonal structure based on sp^2 hybridized carbon atoms) which is the basic structure of all graphitic carbon materials such as carbon nanotubes ("rolled up" graphene sheets) and graphite (stacked graphene sheets) [64]. Graphene has received increasing attention due to its unique physicochemical properties such as high surface area, excellent conductivity, high mechanical strength, stiffness and elasticity, and very high electrical and thermal conductivity [65]. These properties suggest such graphene could replace materials as when combined with others could also enable several disruptive technologies. For example, the combination of conductivity, transparency, and elasticity can be applied to flexible electronic displays [66].

The production of graphene characterizing its unique properties was firstly reported by Geim and Novoselov in 2004 when the material was isolated using the "Scotch Tape" method and its high charge-carrier mobility was determined [64, 67–70]. Geim and Novoselov were awarded the Nobel Prize in physics in 2010 [64]. However, the term graphene as layers of graphitic materials has been reported decades previous to 2004 [68]. In 1958, Hummers and Offeman already have introduced a procedure that is one of the simplest, safe, large-scale, and widely used methods where the base that consists in the obtaining of graphitic oxides from oxidizing in strong acidic media using a mixture of sulfuric and nitric acids with sodium nitrate is used up to nowadays [71].

The "Hummers method" has been improved by replacing the use of $NaNO_3$ with H_2SO_4, H_3PO_4 and double the amount of $KMnO_4$ to avoid the emission of toxic gases caused by the oxidation of $NaNO_3$ (NO_2 e N_2O) [72]. This method produces oxidized form the graphene oxides (GO) that has a large number of oxy-functional groups over the GO surface such as -OH, -COOH, -O, and -CO, makes the material readily dispersible in water and in some polar organic solvents, which promotes also strong interaction with functional groups from monomers such as aniline in an aqueous system [73]. On the other hand, GO loses its conductivity due to changes of carbon–carbon bonds hybridizations from sp^2 to sp^3. In this sense, aiming the improvement of conductivity, new environmentally friendly reducing agents have been used for rGO obtaining [74]. As for example, ascorbic acid emerges as a green alternative to common reducing agents and can be compared to hydrazine for the deoxygenation of GO being safe and innocuous [75].

The properties of graphene are intrinsically dependent of the production method that can affect directly the quality, dimensions, shapes, type of defects, etc. Aiming large-scale industrial applications, several methods for production have been exploited such as ion chemical vapor deposition (CVD) [76], liquid-phase exfoliation [77], growth upon a silicon carbide substrate [78], unzipping of carbon nanotubes [79], among others [66]. The different production methods result in different variations of graphene with different properties and its applications are driven according to its characteristics. In general, graphene or rGO flakes are used for composite materials, conductive paints, and reinforce materials, while planar graphene for lower-performance active and non-active devices and planar graphene for high-performance electronic devices [66].

Graphene with its planar structure, high specific area, high conductivity, and low thickness shows several advantages in the development of sensor and biosensors. One of the improvements, when compared to carbon nanotubes, is that planar graphene sheets can expose all carbon atoms to interact directly with the analytes. Moreover, graphene does not contain metallic impurities that can interfere in the analysis and some production methods are cheaper, large scale, and cost effective [80].

The development of sensors and biosensors which based on different forms of graphene (such as GO and rGO) has been used in several applications in different areas especially clinical (deoxyribonucleic acid (DNA), glucose, cholesterol, H_2O_2, uric acid, ascorbic acid, dopamine, proteins, etc.), environmental (pesticides, metallic ions, hormones, bisphenol A, etc.), and food applications [80]. As for example, the direct electrochemistry of enzyme, its electrocatalytic activity toward small biomolecules (hydrogen peroxide, NADH, dopamine, etc.), and graphene-based enzyme biosensors have been reported for several works [65]. The combination of graphene sheets with nanoparticles or other nanostructured materials can also enhance the analytical response such as sensitivity, the limit of detection, and reproducibility of sensors and biosensors [80].

3.2 Applications of Carbon-Based Materials in Biosensors

Uric acid (UA) is the final product of the metabolism of purines and heterocyclic organic compounds present in the human body. The monitoring of UA in body fluids is of great importance as high concentrations of UA are related to increased health problems and increases the individual's risk of developing kidney disease, heart disease, and diabetes. Disposable biosensor is an excellent alternative to be used in clinical diagnoses involving UA, as they allow a quick and low-cost response. Thus, Shi et al. (2020) [81] developed a disposable electrochemical biosensor using MWCNT to detect UA in human saliva. The authors developed an AutoCAD template of the printed circuit electrode (SPE) and used it on previously cleaned polyethylene terephthalate (PET) substrate. By printing Ag/AgCl conductive ink and carbon conductive ink, reference electrodes and working electrodes/counter electrodes were obtained, respectively. After drying, the working electrode was modified with MWCNT by depositing 3 μL of the aqueous suspension of MWCNT and with 3 μL of the aqueous solution of uricase, which was incubated at 4 °C for 2 h. Uricase is an enzyme found in many organisms, but not in humans. For comparison, the same electrode was manufactured without modification with MWCNT. And both were analyzed in $K_3Fe(CN)_6/K_4Fe(CN)_6$ system. Human saliva samples were collected, following ethical protocols. The samples were subjected to standard addition. The schematic in Fig. 4 (Ia) shows the steps of fabrication of the SPE on PET substrate, including its modification by drop-casting with MWCNT and with the enzyme, Fig. 4 (Ib) shows the final appearance of the SPE, followed by scanning electron microscope (SEM) results (Figs. 4II).

The SEM results show the morphological characterization of the electrodes, where in Figs. 4IIa and IIb it is observed the roughness of the SPE manufactured in 4IIc, the modification with MWCNT and in 4IId the enzyme incubated on the MWCNT/SPE. In the Fig. 4IId, it is possible to see that the uricase incubation allows the electrode surface to become smoother. The CV results showed that modification with MWCNT increases the peak current intensity, but when modifying with the enzyme the current intensity was reduced. This was corroborated by the EIS results, which showed an obvious decrease in R_{ct} for the MWCNT modification which increased again when analyzing the device with the incubated enzyme. The DPV results an oxidation peak at 0.31 V in both devices. However, the modified SPE with MWCNT accentuated the oxidation peak 26 times greater than the modified unmodified SPE, showing that the device increased when modified with MWCNT. The standard addition results also showed a linear increase in current with increasing UA concentration, with a low detection limit. In chrono-amperometry (CA) results, biosensors with MWCNT had a higher response signal. Thus, it was advantageous to modify the SPE with MWCNT, as it increased the sensitivity of the device, allowing it to be used in human saliva [81].

Cancer is known for its degree of lethality and is estimated that there are about 14.1 million malignant cases that lead to 8.2 million deaths annually [82]. By using a biosensor to detect the disease directly in bodily fluids, such as serum and urine,

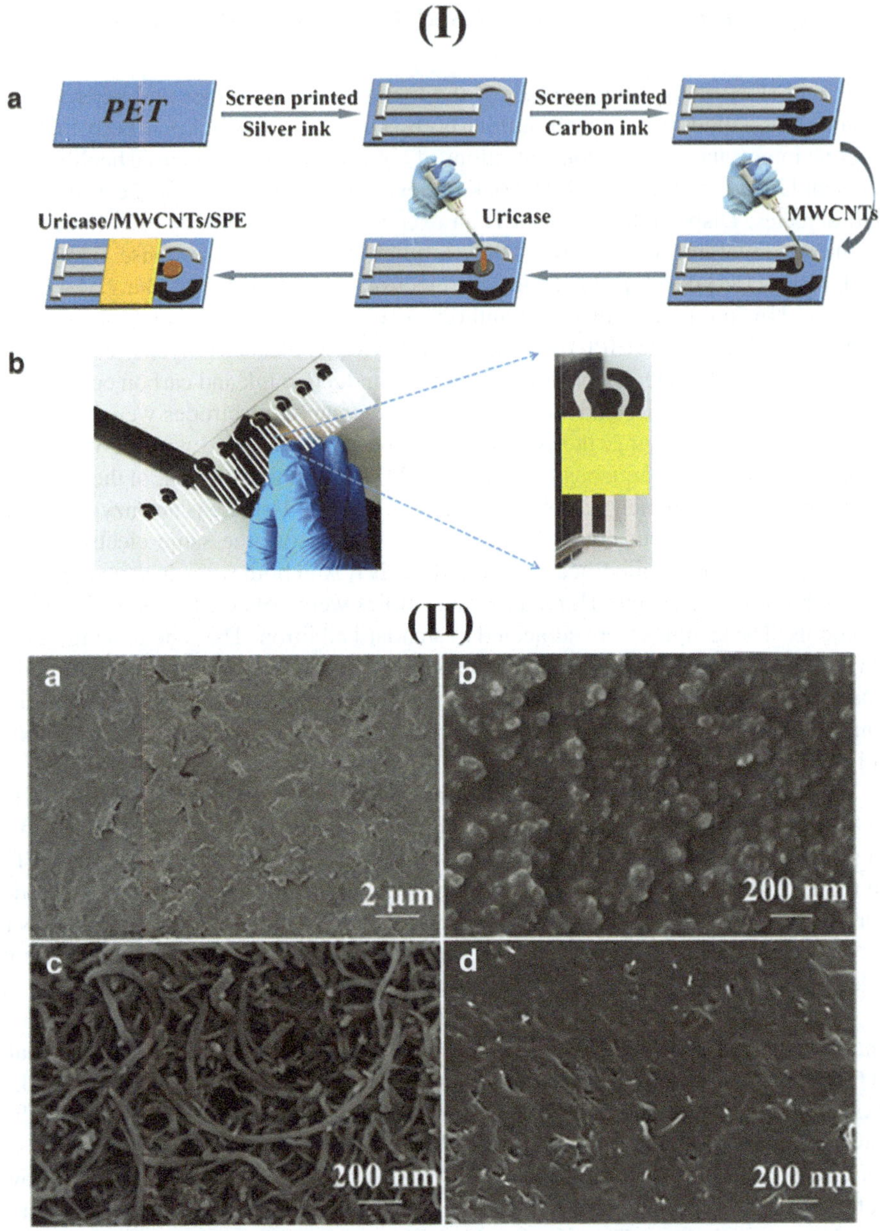

Fig. 4 **Ia** Schematic illustration of the stepwise SPE fabrication; **Ib** images of the final device obtained; **IIa** and **IIb** SEM images of the SPE without modification; **IIc** SEM image of the modification with MWCNT; and **IId** SEM image of morphology of working electrode modified with MWCNT and with the incubated enzyme [81]

non-invasive analyzes can be performed, facilitating the early detection of the disease and favoring its treatment. A carbohydrate antigen is expressed on carcinoma cells which have changes in glycolysis patterns, so the identification of antigens in the expression of carbohydrates allows to identify the progress of the disease. In this work, a graphene biosensor was developed for cancer biomarker detection of the tumor-associated Thomsen-nouvelle (Tn) antigens. The biosensor was developed using graphene screen-printed electrodes (SPE) activated by CA at three time intervals and with two potential values using Ag/AgCl reference electrode and auxiliary platinum electrode for deposition of oxygenated groups such as—COOH. In sequence, the sensor was chemically activated with a solution of 40 μL of N-(3-dimethylaminopropyl)—N'-ethylcarbodiimide (EDC) 200 mM hydrochloride and in 50 mM N-hydroxysuccinimide (NHS) solution. Then, the electrode surface was incubated with human serum albumin (HSA) for covalent correction of HSA as a nanoscaffold, activated with a 40 μL solution of 200 mM EDC and 50 mM NHS. In this way, it was possible to carry out electrochemical oxidation on the electrode and modify it with HSA, and to then covalently immobilize the Tn antigen. After immobilizing the antigen, the analyte was incubated in obligatory up to 9 pM. After optimizing the parameters of electrode activation and modification with HSA, electrochemical characterization using CV was performed, where covalent immobilization of HSA resulted in a decrease in ΔE. The results of the glycan biosensor indicated that it detects the analyte with good sensitivity and selectivity, with satisfactory reproducibility and an excellent detection limit [82].

In addition to cancer, graphene biosensors are interesting alternatives for other pathogens, such as infection caused by the Zika virus. This type of infection is caused by the transmission of the virus, and several outbreaks have taken place in recent history. The infection can cause Guillain–Barre syndrome in adults, and in fetuses, it can cause severe brain damage. A graphene-based biosensor was produced immobilizing a highly specific monoclonal antibody. Graphene films were previously produced by plasma-intensified photolithography, used to passivate and standardize graphene, forming graphene films, which were later encapsulated. The biosensors were functionalized by immobilizing the protein with anti-Zika monoclonal antibodies developed by the Centers for Disease Control. This antibody was diluted in a PBS solution pH = 7.4, with incubation carried out at room temperature. Possible residual groups were extinguished. The biosensor was characterized by AFM and Raman spectroscopy and was applied to analysis in PBS solution (pH = 7.4) and in artificial serum. The results showed that protein immobilization occurs at a density of 5 proteins/μm^2 on the graphene surface with a thickness of 1.0 nm and that during the procedure, in which the graphene is initially functionalized, and then the antibody is immobilized, it results in an increase of 20% in capacitance response. The detection sensitivity was also determined, and the selectivity was validated, showing that the biosensor can be applied in tests to diagnose the Zika virus [83].

A SPE biosensor based on combined graphene and reduced graphene oxide (rGO) was used to determine the beta-amyloid (Aβ) peptide, which is an Alzheimer's biomarker, as anomalous levels of the biomarker are related to pathologies caused

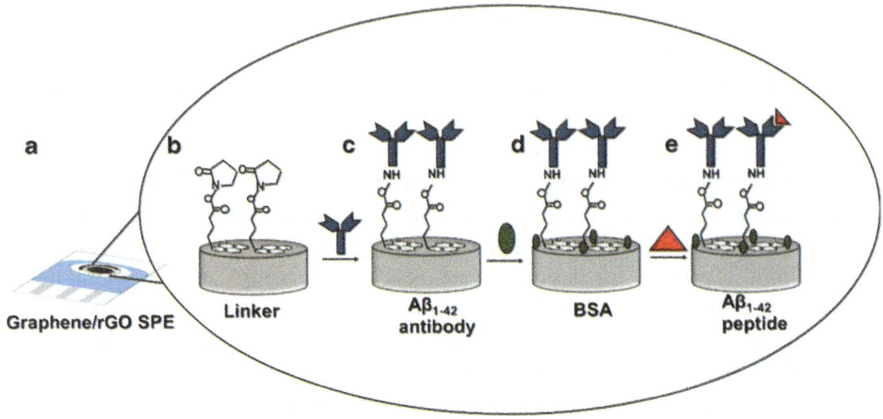

Fig. 5 Schematic of the electrochemical process to detect $A\beta_{1-42}$: SPE graphene/rGO (**a**); modification with the linker (Pyr-NHS) (**b**); antibody (**c**); BSA (**d**); $A\beta_{1-42}$ peptide(**e**) [84]

by Alzheimer's disease. According to the authors, the graphene-coupled rGO generates high conductivity and active sites that can be used in biosensors through the chemical functionalization of rGO, which starts to act as an anchoring site for the antibodies. The biosensor was validated in human plasma samples and in mouse plasma samples. Initially, the SPE was produced by deposition of an aqueous solution of GO on the graphene of the SPE. The GO was electrochemically reduced in 10 mM $K_3[Fe(CN)_6]$ with a 1 M KCl solution, with a CV cycle, varying the potential from −1.5 V to 0.5 V. In sequence, the device was incubated through modification of N-hydroxysuccinimide 1-pyrenebutyric acid (Pyr-NHS) which was performed in methanol to facilitate the immobilization of the Aβ antibody for 2 h and then incubated for 16 h in a solution containing H31L21 antibody. The plasma samples were enriched with the peptide concentration and 20 μL was deposited on the biosensor for 1 h. Then, the electrode surface was blocked with BSA (in PBS). The process performed is summarized in Fig. 5 [84].

The cyclic voltammograms showed that modification with GO reduces the peak current, but an electrochemical reduction of GO promotes the formation of the graphene/rGO film which has a higher peak current intensity than graphene SPE without the modification, due to high conductivity of graphene combined with the electroactive sites of rGO. Furthermore, antibody incubation time and ligand concentration were optimized. Device sensitivity was determined using DPV and the peak current intensity decreased by increasing the Aβ concentration. Limit of detection (LOD) of 2.398 pM was obtained and a linear range from 11 pM to 55 nM. Thus, this graphene and rGO biosensor proved to be an interesting device for the electrochemical detection of the Aβ peptide, reconciling the electrochemical inertia and electrocatalytic activity of graphene with the electroactive sites of rGO.

4 Final Considerations

CPs such as PAni and PPy and carbon-based materials mainly fullerene, carbon nanotubes, and graphene have stood out in bioelectrochemistry due to their characteristics. In this chapter, it was observed that the CPs have been extensively used as active transducer materials in the development of electrodes aiming biosensor applications due to their electrical conductivity can be modulated on a wide range, besides their flexibility, processability, and mechanical properties. Already carbon-based materials been widely used in biosensor applications due to advances in nanoscience and nanotechnology, such as miniaturization of system, less energy consumption, lower cost, and improved functionalities compared to other materials.

Furthermore, the applications discussed were in areas such as clinical and environmental. The clinical applications of CPs-based biosensors and carbon-based materials showed the monitoring of uric acid in body fluids, detecting cancer biomarkers, facilitating the early detection of the disease, and favoring its treatment. In the environmental area, biosensors based on materials discussed throughout this chapter were applied in the detection of emerging pollutants such as pesticide OPs, preventing their accumulation in the environment, in addition to helping to maintain well-being and life.

The main advantages of CPs-based biosensors are their sensitivity and specificity, due to the addition of functional groups or molecules immobilized in their structures, mainly as nanofibers and nanostructured films. Biosensors based on carbon-based materials, on the other hand, also have advantages related to increased sensitivity and selectivity and enable good reproducibility of the device. In this way, it is possible to unite the advantages of these two material classes in hybrids.

The next chapter will address the characteristics and the advantages of the hybrids of intrinsic CPs and carbon-based materials. It will also address the recent applications of these hybrids in bioelectrochemistry, with a focus on biosensors.

References

1. Maduraiveeran, G., Sasidharan, M., Ganesan, V.: Electrochemical sensor and biosensor platforms based on advanced nanomaterials for biological and biomedical applications. Biosens. Bioelectron. **103**, 113–129 (2018)
2. Garcia-Carmona, L., Martin, A., Sempionatto, J.R., et al.: Pacifier biosensor: toward noninvasive saliva biomarker monitoring. Anal. Chem. **91**, 13883–13891 (2019)
3. Seo, G., Lee, G., Kim, M.J., et al.: Rapid detection of COVID-19 causative virus (SARS-CoV-2) in human nasopharyngeal swab specimens using field-effect transistor-based biosensor. ACS Nano **14**, 5135–5142 (2020)
4. Bravo, I., Revenga-Parra, M., Pariente, F., Lorenzo, E.: Reagent-less and robust biosensor for direct determination of lactate in food samples. Sensors **17**, 144 (2017)
5. Gaudin, V.: Advances in biosensor development for the screening of antibiotic residues in food products of animal origin—a comprehensive review. Biosens. Bioelectron. **90**, 363–377 (2017)

6. da Silva, M.K.L., Vanzela, H.C., Defavari, L.M., Cesarino, I.: Determination of carbamate pesticide in food using a biosensor based on reduced graphene oxide and acetylcholinesterase enzyme. Sens.S Actuators B-Chem. **277**, 555–561 (2018)
7. Maduraiveeran, G., Jin, W.: Nanomaterials based electrochemical sensor and biosensor platforms for environmental applications. Trends Environ. Anal. Chem. **13**, 10–23 (2017)
8. Uniyal, S., Sharma, R.K.: Technological advancement in electrochemical biosensor based detection of Organophosphate pesticide chlorpyrifos in the environment: A review of status and prospects. Biosens. Bioelectron. **116**, 37–50 (2018)
9. Perumal, V., Hashim, U.: Advances in biosensors: Principle, architecture and applications. J. Appl. Biomed. **12**, 1–15 (2014)
10. Xu, N., Bai, J., Peng, Y., et al.: Pretreatment-free detection of diazepam in beverages based on a thermometric biosensor. Sens.S Actuators B-Chem. **241**, 504–512 (2017)
11. Shafiee, H., Lidstone, E.A., Jahangir, M., et al.: Nanostructured Optical Photonic Crystal Biosensor for HIV Viral Load Measurement. Sci. Rep. **4**, 4116 (2014)
12. Pohanka, M.: Piezoelectric biosensor for the determination of Tumor Necrosis Factor Alpha. Talanta **178**, 970–973 (2018)
13. Jayanthi, V.S.P.K.S.A., Das, A.B., Saxena, U.: Recent advances in biosensor development for the detection of cancer biomarkers. Biosens. Bioelectron. **91**, 15–23 (2017)
14. Mishra, R.K., Hubble, L.J., Martin, A., et al.: Wearable flexible and stretchable glove biosensor for on-site detection of organophosphorus chemical threats. Acs Sensors **2**, 553–561 (2017)
15. Cui, H.-F., Wu, W.-W., Li, M.-M., et al.: A highly stable acetylcholinesterase biosensor based on chitosan-TiO(2)graphene nanocomposites for detection of organophosphate pesticides. Biosens. Bioelectron. **99**, 223–229 (2018)
16. Martinkova, P., Pohanka, M.: Biosensors for blood glucose and diabetes diagnosis: evolution, construction, and current status. Anal. Lett. **48**, 2509–2532 (2015)
17. Thevenot, D.R., Toth, K., Durst, R.A., Wilson, G.S.: Electrochemical biosensors: recommended definitions and classification. Biosens. Bioelectron. **16**, 121–131 (2001)
18. Clark, L., Lyons, C.: Electrode systems for continuous monitoring in cardiovascular surgery. Ann. N. Y. Acad. Sci. **102**, 29–000 (1962)
19. Shirakawa, H., Louis, E.J., MacDiarmid, A.G., et al.: Synthesis of electrically conducting organic polymers: halogen derivatives of polyacetylene, (CH) x. J. Chem. Soc., Chem. Commun. **16**, 578–580 (1977)
20. Aqel, A., Abou El-Nour, K.M.M., Ammar, R.A.A., Al-Warthan, A.: Carbon nanotubes, science and technology part (I) structure, synthesis and characterisation. Arab. J. Chem. **5**, 1–23 (2012)
21. Li, Y.: Conducting polymers. In: Organic Optoelectronic Materials. Springer, pp. 23–50 (2015)
22. Pron, A., Rannou, P.: Processible conjugated polymers: from organic semiconductors to organic metals and superconductors. Prog. Polym. Sci. **27**, 135–190 (2002)
23. Chiang, J.C., Macdiarmid, A.G.: Polyaniline—protonic acid doping of the emeraldine form to the metallic regime. Synth. Met. **13**, 193–205 (1986)
24. Levi, M.D., Vorotyntsev, M.A., Kazarinov, V.E., Frumkin, A.N.: Electron-conducting polymers—electrochemical doping and equilibrium potential distribution across metal-polymer-solution interfaces. Synth. Met. **43**, 2923–2925 (1991)
25. Macdiarmid, A.G., Epstein, A.J.: Secondary doping in polyaniline. Synth. Met. **69**, 85–92 (1995)
26. Paasch, G.: Transmission line description for doped conjugated polymers with polarons, bipolarons and counterions as charged species. Electrochim. Acta **47**, 2049–2053 (2002)
27. Paasch, G.: Transport in doped conjugated polymers with polarons and bipolarons forming complexes with counter ions. Solid State Ion.S **169**, 87–94 (2004)
28. Popkirov, G.S., Barsoukov, E.: In-situ impedance measurements during oxidation and reduction of conducting polymers—significance of the capacitive currents. J. Electroanal. Chem. **383**, 155–160 (1995)
29. Lange, U., Roznyatouskaya, N.V., Mirsky, V.M.: Conducting polymers in chemical sensors and arrays. Anal. Chim. Acta **614**, 1–26 (2008)

30. Bartlett, P., Birkin, P.: The application of conducting polymers in biosensors. Synth. Met. **61**, 15–21 (1993)
31. Xia, L., Wei, Z., Wan, M.: Conducting polymer nanostructures and their application in biosensors. J. Colloid Interface Sci. **341**, 1–11 (2010)
32. Lee, H.-Y., Jung, Y., Kim, S.: Conducting polymer coated graphene oxide electrode for rechargeable lithium-sulfur batteries. J. Nanosci. Nanotechnol. **16**, 2692–2695 (2016)
33. An, H.F., Wang, X.Y., Li, N., et al.: Carbon and conducting polymer composites for supercapacitors. Prog. Chem. **21**, 1832–1838 (2009)
34. Ghanbari, K., Mousavi, M.F., Shamsipur, M.: Preparation of polyaniline nanofibers and their use as a cathode of aqueous rechargeable batteries. Electrochim. Acta **52**, 1514–1522 (2006)
35. Ebrahim, S., El-Raey, R., Hefnawy, A., et al.: Electrochemical sensor based on polyaniline nanofibers/single wall carbon nanotubes composite for detection of malathion. Synth. Metals **190**, 13–19 (2014)
36. Paterno, L.G., Fonseca, F.J., Alcantara, G.B., et al.: Fabrication and characterization of nanostructured conducting polymer films containing magnetic nanoparticles. Thin Solid Films **517**, 1753–1758 (2009)
37. Kucherenko, I.S., Soldatkin, O.O., Kucherenko, D.Y., et al.: Advances in nanomaterial application in enzyme-based electrochemical biosensors: a review. Nanoscale Advances **1**, 4560–4577 (2019)
38. Sethuraman, V., Muthuraja, P., Raj, J.A., Manisankar, P.: A highly sensitive electrochemical biosensor for catechol using conducting polymer reduced graphene oxide-metal oxide enzyme modified electrode. Biosens. Bioelectron. **84**, 112–119 (2016)
39. Dai, H., Wang, N., Wang, D., et al.: Voltammetric uric acid sensor based on a glassy carbon electrode modified with a nanocomposite consisting of polytetraphenylporphyrin, polypyrrole, and graphene oxide. Microchim. Acta **183**, 3053–3059 (2016)
40. Gerard, M., Chaubey, A., Malhotra, B.D.: Application of conducting polymers to biosensors. Biosens.S & Bioelectron. **17**, 345–359 (2002)
41. Zare, E.N., Makvandi, P., Ashtari, B., et al.: Progress in conductive polyaniline-based nanocomposites for biomedical applications: a review. J. Med. Chem. **63**, 1–22 (2020)
42. Lai, J., Yi, Y., Zhu, P., et al.: Polyaniline-based glucose biosensor: a review. J. Electroanal. Chem. **782**, 138–153 (2016)
43. Hamid, H.H., Harb, M.E., Elshaer, A.M., et al.: Electrochemical preparation and electrical characterization of polyaniline as a sensitive biosensor. Microsyst. Technol. Micro Nanosyst. -Inf. Storage Process. Syst. **24**, 1775–1781 (2018)
44. Asmatulu, R., Veisi, Z., Uddin, M.N., Mahapatro, A.: Highly sensitive and reliable electrospun polyaniline nanofiber based biosensor as a robust platform for COX-2 enzyme detections. Fibers Polym. **20**, 966–974 (2019)
45. Mishra, R.K., Barfidokht, A., Karajic, A., et al.: Wearable potentiometric tattoo biosensor for on-body detection of G-type nerve agents simulants. Sens.S Actuators B-Chem. **273**, 966–972 (2018)
46. Wang, H., Ma, Z.: A cascade reaction signal-amplified amperometric immunosensor platform for ultrasensitive detection of tumour marker. Sens.S Actuators B-Chem. **254**, 642–647 (2018)
47. Sheikhzadeh, E., Chamsaz, M., Turner, A.P.F., et al.: Label-free impedimetric biosensor for Salmonella Typhimurium detection based on poly [pyrrole-co-3-carboxyl-pyrrole] copolymer supported aptamer. Biosens. Bioelectron. **80**, 194–200 (2016)
48. Li, S., Jiang, Y., Eda, S., Wu, J.J.: Low-cost and desktop-fabricated biosensor for rapid and sensitive detection of circulating D-dimer biomarker. IEEE Sens. J. **19**, 1245–1251 (2019)
49. Ayenimo, J.G., Adeloju, S.B.: Amperometric detection of glucose in fruit juices with polypyrrole-based biosensor with an integrated permselective layer for exclusion of interferences. Food Chem. **229**, 127–135 (2017)
50. Akdag, A., Isik, M., Goktas, H.: Conducting polymer-based electrochemical biosensor for the detection of acetylthiocholine and pesticide via acetylcholinesterase. Biotechnol. Appl. Biochem **68**, 1113–1119 (2021)

51. Bhong, S.Y., More, N., Choppadandi, M., Kapusetti, G.: Review on carbon nanomaterials as typical candidates for orthopaedic coatings. Sn Appl. Sci. **1**, 76 (2019)
52. Stamatin, I., Morozan, A., Dumitru, A., et al.: The synthesis of multi-walled carbon nanotubes (MWNTs) by catalytic pyrolysis of the phenol-formaldehyde resins. Phys. E-Low-Dimens. Syst. & Nanostructures **37**, 44–48 (2007)
53. Thostenson, E.T., Ren, Z.F., Chou, T.W.: Advances in the science and technology of carbon nanotubes and their composites: a review. Compos. Sci. Technol. **61**, 1899–1912 (2001)
54. Iijima, S.: Helical microtubules of graphitic carbon. Nature **354**, 56–58 (1991)
55. Andrews, R., Weisenberger, M.C.: Carbon nanotube polymer composites. Curr. Opin. Solid State Mater. Sci. **8**, 31–37 (2004)
56. Dai, H.J.: Carbon nanotubes: synthesis, integration, and properties. Acc. Chem. Res. **35**, 1035–1044 (2002)
57. Hazani, M., Hennrich, F., Kappes, M., et al.: DNA-mediated self-assembly of carbon nanotube-based electronic devices. Chem. Phys. Lett. **391**, 389–392 (2004)
58. He, B.L., Dong, B., Wang, W., Li, H.L.: Performance of polyaniline/multi-walled carbon nanotubes composites as cathode for rechargeable lithium batteries. Mater. Chem. Phys. **114**, 371–375 (2009)
59. Qiu, Y.J., Yu, J., Fang, G., et al.: Synthesis of carbon/carbon core/shell nanotubes with a high specific surface area. J. Phys. Chem. C **113**, 61–68 (2009)
60. Zhao, Q., Gan, Z., Zhuang, Q.: Electrochemical sensors based on carbon nanotubes. Electroanal. **14**, 1609–1613 (2002)
61. Hwang, H.S., Jeong, J.W., Kim, Y.A., Chang, M.: Carbon nanomaterials as versatile platforms for biosensing applications. Micromachines **11**, 814 (2020)
62. Rivas, G.A., Rubianes, M.D., Rodriguez, M.C., et al.: Carbon nanotubes for electrochemical biosensing. Talanta **74**, 291–307 (2007)
63. Valentini, F., Orlanducci, S., Terranova, M.L., et al.: Carbon nanotubes as electrode materials for the assembling of new electrochemical biosensors. Sens.S Actuators B-Chem. **100**, 117–125 (2004)
64. Geim, A.K., Novoselov, K.S.: The rise of graphene. Nature Mater **6**, 183–191 (2007)
65. Shao, Y., Wang, J., Wu, H., et al.: Graphene based electrochemical sensors and biosensors: a review. Electroanalysis **22**, 1027–1036 (2010)
66. Novoselov, K.S., Fal′ko, V.I., Colombo, L., et al.: A roadmap for graphene. Nature **490**, 192–200 (2012)
67. Novoselov, K.S., Geim, A.K., Morozov, S.V., et al.: Electric field effect in atomically thin carbon films. Science **306**, 666 (2004)
68. Novoselov, K.S., Jiang, D., Schedin, F., et al.: Two-dimensional atomic crystals. Proc. Natl. Acad. Sci. **102**, 10451–10453 (2005)
69. Novoselov, K.S., Geim, A.K., Morozov, S.V., et al.: Two-dimensional gas of massless Dirac fermions in graphene. Nature **438**, 197–200 (2005)
70. Zhang, Y., Tan, Y.-W., Stormer, H.L., Kim, P.: Experimental observation of the quantum Hall effect and Berry's phase in graphene. Nature **438**, 201–204 (2005)
71. Hummers, W.S., Offeman, R.E.: Preparation of graphitic oxide. J. Am Chem. Soc. **80**, 1339–1339 (1958)
72. Zaaba, N.I., Foo, K.L., Hashim, U., et al.: Synthesis of graphene oxide using modified hummers method: solvent influence. Procedia Eng. **184**, 469–477 (2017)
73. Li, X., Zhong, Q., Zhang, X., et al.: In-situ polymerization of polyaniline on the surface of graphene oxide for high electrochemical capacitance. Thin Solid Films **584**, 348–352 (2015)
74. Zhang, J., Yang, H., Shen, G., et al.: Reduction of graphene oxide via L-ascorbic acid. Chem. Commun. **46**, 1112–1114 (2010)
75. Xu, C., Shi, X., Ji, A., et al.: Fabrication and characteristics of reduced graphene oxide produced with different green reductants. Plos One **10**, e0144842 (2015)
76. Losurdo, M., Giangregorio, M.M., Capezzuto, P., Bruno, G.: Graphene CVD growth on copper and nickel: role of hydrogen in kinetics and structure. Phys. Chem. Chem. Phys. **13**, 20836 (2011)

77. Hernandez, Y., Nicolosi, V., Lotya, M., et al.: High-yield production of graphene by liquid-phase exfoliation of graphite. Nat. Nanotech **3**, 563–568 (2008)
78. de Heer, W.A., Berger, C., Ruan, M., et al.: Large area and structured epitaxial graphene produced by confinement controlled sublimation of silicon carbide. Proc. Natl. Acad. Sci. **108**, 16900–16905 (2011)
79. Kosynkin, D.V., Higginbotham, A.L., Sinitskii, A., et al.: Longitudinal unzipping of carbon nanotubes to form graphene nanoribbons. Nature **458**, 872–876 (2009)
80. Justino, C.I.L., Gomes, A.R., Freitas, A.C., et al.: Graphene based sensors and biosensors. TrAC, Trends Anal. Chem. **91**, 53–66 (2017)
81. Shi, W., Li, J., Wu, J., et al.: An electrochemical biosensor based on multi-wall carbon nanotube-modified screen-printed electrode immobilized by uricase for the detection of salivary uric acid. Anal. Bioanal. Chem. **412**, 7275–7283 (2020)
82. Kveton, F., Blsakova, A., Lorencova, L., et al.: A graphene-based glycan biosensor for electrochemical label-free detection of a tumor-associated antibody. Sensors **19**, 5409 (2019)
83. Afsahi, S., Lerner, M.B., Goldstein, J.M., et al.: Novel graphene-based biosensor for early detection of Zika virus infection. Biosens. Bioelectron. **100**, 85–88 (2018)
84. Sethi, J., Van Bulck, M., Suhail, A., et al.: A label-free biosensor based on graphene and reduced graphene oxide dual-layer for electrochemical determination of beta-amyloid biomarkers. Microchim. Acta **187**, 288 (2020)

3D-Printed Electrochemical Devices for Sensing and Biosensing of Biomarkers

Luiz R. G. Silva, Ava Gevaerd, Luiz H. Marcolino-Junior, Márcio F. Bergamini, Tiago Almeida Silva, and Bruno Campos Janegitz

Abstract In this chapter, we explore the concepts of obtaining 3D devices, as well as the most relevant results regarding applications of these electrochemical sensors/biosensors for the determination of biomarkers. Some procedures and trends, new cell geometries, surface treatments, and applications in fully printed tools are described. Furthermore, the applications of the devices for biomarkers determination and their advantages will be discussed in detail, as well as the perspectives for the use of these devices.

Keywords 3D printing · Electrochemistry · Biomarkers · Sensors · Biosensors

1 Introduction

The electrochemical sensors obtained by 2D-printing techniques are widely used for the monitoring of biomarkers in general, and their surface can be modified with different types of biomolecules, or treatments depending on the final application purpose. However, some limitations on the sensor geometry, on the inks used in the process of obtaining the strips, in addition to limitations on the 2D-printing methodology, can restrict the scope of application of these sensors for biological applications [1, 2]. Thus, as a way to get around and even improve performance, 3D printing comes as a promising approach, allowing the devices obtained by this

L. R. G. Silva · B. C. Janegitz (✉)
Laboratory of Sensors, Nanomedicine, and Nanostructured Materials, Federal University of São Carlos, Araras, São Paulo, Brazil
e-mail: brunocj@ufscar.br

A. Gevaerd · L. H. Marcolino-Junior · M. F. Bergamini
Laboratório de Sensores Eletroquímicos (LabSensE), Departamento de Química, Universidade Federal Do Paraná (UFPR), Curitiba, Paraná, Brazil

A. Gevaerd
Hilab, Curitiba, Paraná, Brazil

T. A. Silva
Department of Chemistry, Federal University of Viçosa, Viçosa, Minas Gerais, Brazil

F. N. Crespilho (ed.), *Advances in Bioelectrochemistry Volume 2*,
https://doi.org/10.1007/978-3-030-95270-9_7

methodology to be more robust, stable, and enabling more stable measurements when in biological environments, such as fluids biomarkers in which the biomarkers are found [3].

3D printing is one of the most innovative technologies available today, and it has established itself as a popular and powerful tool in many different fields. If we look at the last 30 years, it is possible to observe the growth of this technology in the most different niches of society [4, 5]. The main motivation in using 3D printing is to obtain customized prototypes for different applications, quickly, with decentralized manufacturing and easy integration with the other parts of the system. This is possible because, in a very simple way, 3D printing is an additive manufacturing method that can build objects from a computational model, a model that is custom designed for the system in question.

In a very recent review, Su [6] makes a survey of 3D analytical devices with biological applications, in which the author concludes that the increasing use and replacement of conventional devices by 3D devices come from a series of characteristics that can be changed when these devices are manufactured, which ultimately generate a result far superior to conventional devices, precisely because of the synergism that is generated when all the best features are placed in one device.

As a brief search in the literature, it is possible to note that a large portion of the application of this technology is for scientific research, and just as 3D printing offers a huge potential for devices from different areas, for systems and research in electroanalytical chemistry would not be different. In the last 10 years, it is possible to note an exponential growth in the publication numbers in the analytical field, especially for the electroanalytical devices (Fig. 1).

For most electroanalytical devices and applications, the 3D devices were obtained using the extrusion of melting thermoplastics, by using fused deposition modeling (FDM) [1, 2]. Although it seems that these devices obtained by FDM have application limitations, that are directly related to the most common thermoplastics, the acrylonitrile butadiene styrene (ABS), and the polylactic acid (PLA), what can be observed, and will be discussed in detail throughout the chapter, is that different types of surface treatments, functionalizations, and changes in the composition of

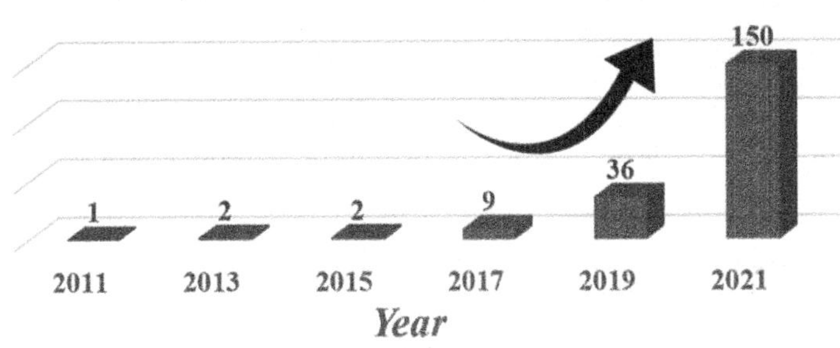

Fig. 1 Publications from the last 10 years in electroanalytical field. Keywords: "3D-printed electrode". Results from Science Direct database

thermoplastics can be carried out, thus improving the interface of these devices with biological environments, making the application range only increase.

As described by Abdalla and Patel [7], electrochemical devices obtained by 3D printing are a new horizon for measurements of biologically relevant molecules, and it is to demonstrate this new horizon that the following discussions were made.

2 3D-Printed Electrochemical Devices

The electrodes used at electrochemical sensors and biosensors are based on solid substrates manufactured by traditional processes. Carbon-based electrodes, especially glassy carbon and carbon paste electrodes, and metallic electrodes such as platinum and gold provide robust surfaces with good electronic conductivity and are versatile for the incorporation of modifiers. Despite the expressive success of the aforementioned working electrodes, currently, there is a new field of research in wide expansion, which aims to manufacture electrode materials and complete electrochemical devices for electroanalysis using a simple, low-cost, and reproducible methodology. Such research makes use of 3D-printing technology, developed at the end of the last century and with significant advances in the current twenty-first century.

3D-printing technology is based on the fabrication of solid objects from the sequential deposition of layers of a given material. In a typical procedure, the first step consists of drawing the desired object in a virtual environment using CAD software, when all dimensions, shapes, and geometries are defined [8]. The file is then converted into a format that is universally recognized and read by the 3D-printer software, the STL (StereoLithography) format [8]. In the printer's software, the 3D model is subjected to a slicing process, which consists of generating several layers of a 2D cross-section of the entire object. Finally, the printer starts to deposit the material after successive sequencing of these 2D layers, which are built on top of each other, until the desired 3D object is fully printed [8]. There are several 3D-printing technologies available, in which the most popular is based on Fused Deposition Modeling (FDM). This technology makes use of an extrusion method, in which a thermoplastic filament is heated to its semi-molten state before extrusion by a movable heated nozzle, which deposits the polymer onto a substrate [1, 9]. Some examples of thermoplastics for use in 3D-FDM printing are polylactic acid (PLA) and acrylonitrile butadiene styrene (ABS). Once deposited, the material solidifies creating a layer that stacks on top of the previous layer. This step is repeated layer-by-layer until the entire object is printed [9].

3D-FDM printers have allowed the dissemination of 3D-printing technology to various branches, due to their lower cost, relatively easy operation, versatility in the design of the most varied and complex objects, and rapid prototyping [10]. Currently, in the context of electrochemistry, 3D-printing technology has already been explored in areas such as the development of Li-ion batteries, capacitors, electrocatalysis, and electroanalysis [8, 11, 12]. In the latter case, 3D printing has been applied in the manufacture of electrodes for three-electrode systems [13]. For this, the conductivity of the used material must be high for electrochemical/electroanalytical applications, and therefore, filaments based on polymers doped with carbonaceous conductive materials (graphene, graphite, and carbon black) are already commercially available, which enabled the manufacture of the electrodes.

In addition to the preparation of the electrodes themselves, an important advance reported is the use of 3D printing for the fabrication of the entire electrochemical device, that are, electrodes and electrochemical cells, the so-called all-in-one 3D-printed electrochemical devices [1]. In the case of the electrochemical cell, to avoid interference with the electrochemical transduction of the three-electrode system, naturally, a non-conductive filament is chosen to manufacture the electrochemical cells. The electrochemical cell and electrodes 3D printed by Richter et al. [14] are displayed in Fig. 2. The components were fabricated by FDM 3D printing using a carbon black/PLA filament for the electrodes and insulant PLA filament for the other electrochemical cell parts.

Although a conductive filament is used, the surface that remains exposed to the electrolytic solution does not yet have fully adequate charge transfer characteristics. This is because conductive particles (such as graphene sheets or carbon black nanoparticles) are still occluded in the insulating polymer matrix. Thus, chemical, electrochemical, and/or mechanical treatment procedures are essential to active the electrodic surface [1]. A systematic study was recently reported by Kalinke et al. [15], in which the use of mechanical polishing, chemical, and electrochemical treatments, individually or in combination, of electrodes printed with PLA/graphene (PLA-G) was explored. The best electrochemical performance was achieved by applying the combined chemical and electrochemical activation steps: (1) chemical treatment by

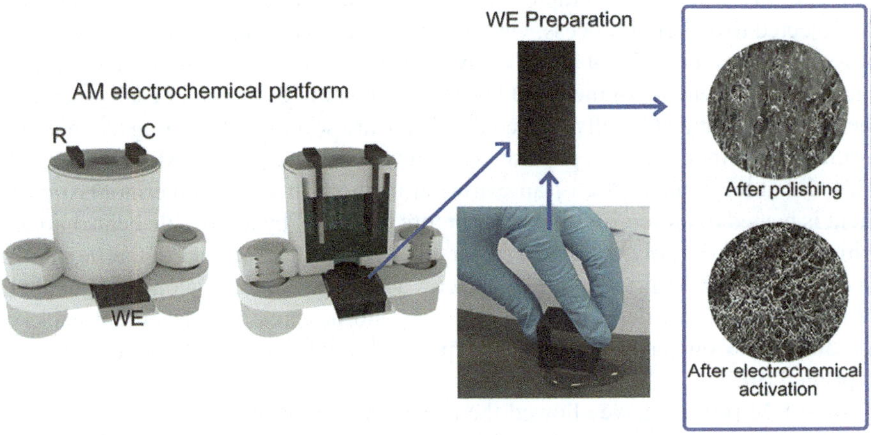

Fig. 2 Schematic 3D-printed (AM) electrochemical cell and the 3D-printing working electrode preparation by polishing. On right, SEM images of the 3D-printed surface after polishing and after electrochemical activation. (Reprinted with permission from [14], Copyright (2019), American Chemical Society)

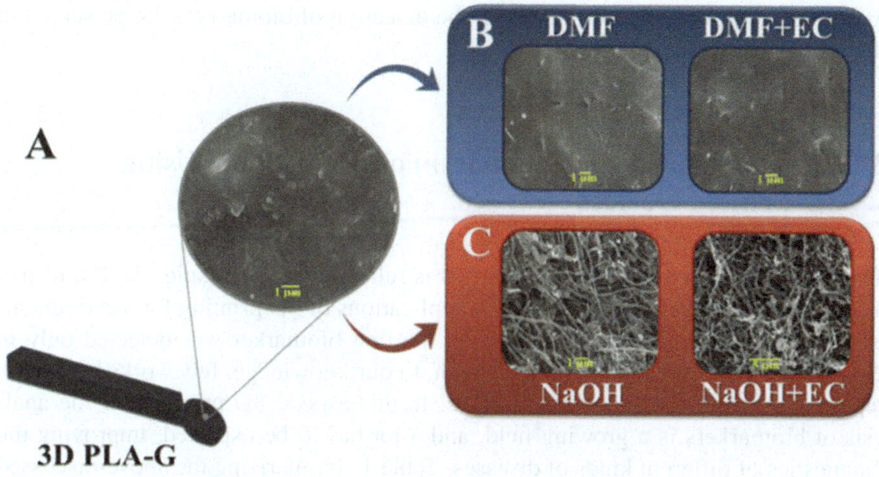

Fig. 3 3D PLA-G electrode design and SEM images with 10 000 × magnification of the electrodes: **a** PLA-G, **b** PLA-G in DMF for 10 min and DMF followed by electrochemical treatment (EC), **c** PLA-G in 1.0 mol L^{-1} NaOH for 30 min and NaOH followed by electrochemical treatment (EC). (Reproduced from [15] with permission from the Royal Society of Chemistry)

immersion in 1.0 mol L^{-1} NaOH solution during 30 min and (2) electrochemical treatment carried out by applying + .8 V during 900 s followed by cyclic voltammetry (potential range of 0.0 to − 1.8 V and scan rate of 50 mV s^{-1}), both in 0.10 mol L^{-1} phosphate buffer solution (pH 7.4). Figure 3 compares the morphology of as-printed electrodes and those obtained from different treatments. Indeed, the SEM images suggest the presence of multilayer graphene nanoribbons, free of the dense PLA layer in the case of the NaOH, and EC-treated electrodes, which led to a great increase in surface area, defects, electron transfer rate, and amount of edge site.

The electrodes printed from the conductive filaments have been applied so far for the determination of analytes of pharmaceutical, environmental, and biological interest. In some cases, analytes with well-explored electrochemical behavior were determined, such as dopamine [10], uric acid [16], nitrite ion [16], heavy metals [17], among others. This is necessary to enable the comparison of data with those provided by electrodes conventionally used in electroanalysis. However, it opens up a great deal of space for the use of this new 3D (bio)sensors for the electrochemical study and quantification of more specific organic and inorganic analytes. 3D–printed electrodes can be used as a conventional non-modified electrode or as a platform for the preparation of electrochemical sensors or biosensors in electroanalysis. In addition to direct use, some recent works propose the modification of electrodes to improve electrochemical properties, such as electrochemically active area and charge transfer kinetics, or biological agents for the proposition of electrochemical biosensors. This modification can be performed by incorporating modifiers in the composition of the filament [18] or by modifying the electrode surface [19, 20]. More examples of application specifically in the detection of biomarkers are presented in the following section.

3 Electrochemical Determination of Biomarkers Using 3D-Printed Devices

The use of 3D printing in electroanalysis is relatively new. Despite the use of this technology for decades, one of the first applications of 3D printing for electrochemistry was demonstrated in 2010 [8], and the first biomarker was detected only in 2018 using 3D-printed sensors. Since then, to our knowledge, few works have been reported for the detection of biomarkers. In this aspect, 3D printing for the analysis of biomarkers is a growing field, and a lot has to be explored, improving the diagnostics of different kinds of diseases. Table 1 summarizing the above-discussed works is presented below, containing the possible biomarker studied, the material used for the fabrication of the sensors, the 3D-printing method employed, as well as the analytical characteristics obtained in each work.

The precursor work employing 3D-printed technology for the detection of a biomarker was reported in 2018 [21]. In this work, a 3D-printed stainless-steel electrode in helical shape was constructed for the detection of paracetamol and dopamine (DA). The 3D-printed stainless-steel electrode was surface-modified with a thin gold layer by electro-plating by applying a constant current (−20 mA for 90 min) to improve the electrochemical performance. The electrodes were obtained using selective laser melting SLM, based on the application of a focused laser beam of high energy which binds metallic particles deposited in a powder form in a printing stage, forming a previously established design, layer by layer. Figure 4 presents an illustration of the design and 3D-printed electrodes, including the Au-modified stainless-steel electrode. The simultaneous detection of the analytes was successfully performed, demonstrating that the 3D-printed sensor is capable of detecting both without mutual interference. Following the same approach, Ho et al. used the Au-modified 3D-printed stainless-steel electrode for the detection of the biomarkers ascorbic acid (AA) and uric acid (UA) [22].

Regarding the use of commercial conductive filaments, Santos et al. [23] developed inexpensive and reproducible 3D-printed graphene electrodes employing the FDM printing technique with conductive polylactic acid/graphene (PLA/graphene) filaments for electrocatalytic detection of dopamine (DA). The electrodes were 3D printed using an extrusion temperature of 190 °C in the form of disks with diameters of 5 mm and thickness of 1 mm. A 2–mm-thick strip was designed to allow the electrode to be connected to a connector. An illustration of the working electrode is shown in Fig. 5.

Still employing the FDM technique, an interesting work that presents a way to manufacture the full electrochemical assembly (electrochemical cell and electrodes) in a single printing step was presented by Katseli et al. [24]. Its applicability was tested

Table 1 Analytical features of 3D electrochemical (bio)sensors toward biomarkers determination

Biomarker	Material	Method	Technique	Concentration range	LOD	Sample	Ref
Dopamine	PLA/graphene	FDM	SWV	5.0–1000 μmol L^{-1}	1.67 μmol L^{-1}	Synthetic human urine and serum	[15]
	PLA/CB	FDM	SWV	1–250 μmol L^{-1}	0.1 μmol L^{-1}	PBS (pH 7.2)	[14]
	PLA/graphene	FDM	DPV	2.0–93.8 μmol L^{-1}	0.24 μmol L^{-1}	PBS (pH 7.4)	[23]
	Stainless steel	SLM	DPV	50–250 μmol L^{-1}	–	PBS (pH 7.4)	[21]
Ascorbic acid	Stainless steel	SLM	DPV	0.1–1.0 mmol L^{-1}	2.1 μmol L^{-1}	Vitamin Cf	[22]
Uric acid	Stainless steel	SLM	DPV	0.1–1.0 mmol L^{-1}	84.0 μmol L^{-}	PBS (pH 7.1) and	[22]
	PLA/graphene	FDM	BIA-MPA	0.5–250 μmol L^{-1}	0.02 μmol L^{-1}	Urine and saliva	[16]
L-methionine	PLA/graphene	FDM	SWV	5.0–3000 μmol L^{-1}	1.39 μmol L^{-1}	Serum	[25]
Hydrogen peroxide	PLA/graphene	FDM	Amperometric	25–100 μmol L^{-1}	9.1	Human serum	[27]
	PLA/CB	FDM	Amperometric	1.5–13.5 mmol L^{-1}	–	–	[29]
C-reactive protein	PLA/CB	FDM	SWV	0–50 ng mL^{-1}	0.06 ng mL^{-1}	Artificial blood	[29]
Glucose	PLA/carbon	FDM	Amperometric	2–28 mmol L^{-1}	–	ABS (pH 4.5)	[24]
	PLA/graphene + $Ni(OH)_2$	FDM	Amperometric	75–1000 μmol L^{-1}	2.4 μmol L^{-1}	NaOH	[26]
	PLA/graphene	FDM	Amperometric	0.5–6.3 mmol L^{-1}	0.015 mmol L^{-1}	Blood plasma	[16]
Caffein	PLA/carbon	FDM	DPV	0–90 mg L^{-1}	1.8 mg L^{-1}	ABS (pH 4.5)	[24]
Nitrite	PLA/graphene	FDM	BIA-MPA	0.5–250 μmol L^{-1}	0.03 μmol L^{-1}	Urine and saliva	[16]
Serotonin	PLA/graphene	FDM	DPV	0.30–10.0 μmol L^{-1}	0.032 μmol L^{-1}	Synthetic urine	[28]
Catechol	PLA/graphene	FDM	SWV	30–700 μmol L^{-1}	0.26 μmol L^{-1}	Natural water	[28]
	PLA/graphene	FDM	CV	0.2–5.0 mmol L^{-1}	7.7 μmol L^{-1}	$HClO_4$	[31]

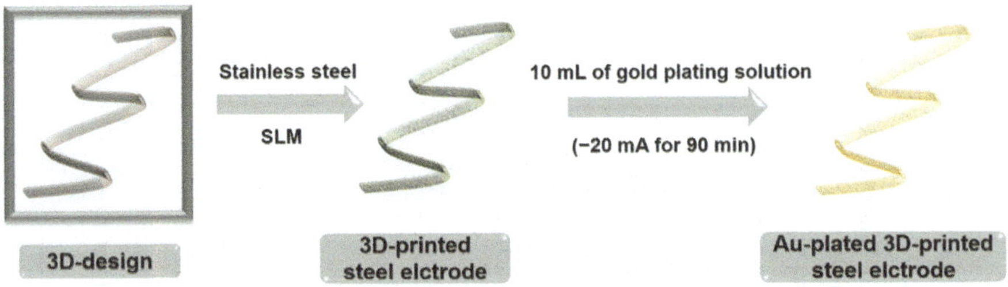

Fig. 4 Schematic representation of the metal 3D-printed electrode fabrication and modification. (Adapted from [21])

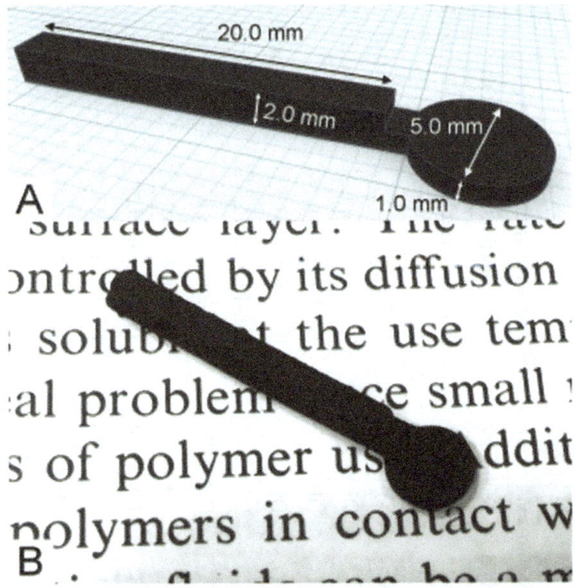

Fig. 5 **a** Design drawn in the Tinkercad platform and **b** the digital photo of the 3D graphene electrode. (Reprinted from [23] with permission of Elsevier)

facing the determination of caffeine and glucose. The three conductive electrodes (working, counter, and pseudo-reference electrodes) were printed from a conductive PLA/carbon filament, and an electrode holder was printed from a non-conductive PLA filament. Figure 6 presents an illustrated diagram of the printer used as well as the final printed electrodes.

Already in 2020, the number of works employing 3D printing for the detection of biomarkers increased. A good example is the work proposed by Kalinke et al. [15], which exploited different types of surface treatments on FDM 3D-printed graphene electrodes for the determination of DA in synthetic urine and serum. The electrodes used in this work also presented a disk shape with 5.0 mm in diameter and 1.0 mm in thickness, and with a connector of 20.0 mm in length and 2.0 mm in thickness. A series

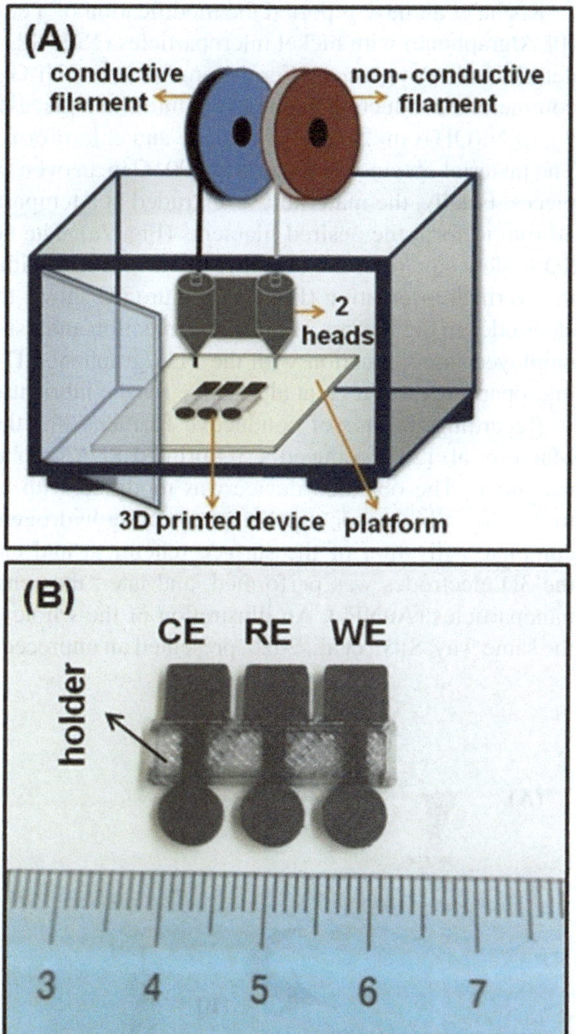

Fig. 6 **a** Schematic illustration of the3D-printing processing of a 3D printer equipped with two heads. **b** Photograph of the 3D-printed integrated device. (Reprinted from [24] with permission of Elsevier)

of surface treatments and its combination were exploited, including the electrochemical activation, mechanical treatment (polishing), and chemical treatments by direct immersion in different types of solvents (dimethylformamide—DMF, NaOH, HNO_3, and H_2SO_4). The optimized surface treatment was a combination of direct immersion in NaOH for 30 min with electrochemical treatment that consist of applying a constant + 1.8 V potential for 900s in the presence of 0.1 mol L^{-1} PBS (pH = 4). The same strategy was used for the determination of L-methionine [25]. The developed sensor was tested in the analysis of biological samples (serum) enriched with L-methionine and presented adequate recovery values, indicating a great potential of the produced 3D-printed sensors for the determination of this biomarker.

Rocha et al. have reported the modification of a commercial conductive filament (PLA/graphene) with nickel microparticles (Ni-G-PLA) [26] for the non-enzymatic determination of glucose. For the production of Ni-G-PLA, a mixture of 30 g of the commercial conductive filament (cut into small pieces) was solubilized together with 3 g of $Ni(OH)_2$ in 250 mL of acetone and chloroform compound solution (3:1 v/v). The material obtained was dried at 100 °C in an oven for 12 h and then cut into small pieces. Finally, the material was extruded at a temperature of 220°C and a speed of 30 rpm to form the desired filaments (Fig. 7a). The new filament was used to print 3D hollow square boxes (4 cm × 4 cm × 2 cm) with a wall thickness of 0.72 mm in a vertical orientation (Fig. 7b). Figure 6C shows the coupling of the 3D-printed electrodes to the bottom of the batch injection analysis cell (BIA). Thus, the method employed in conjunction with the PLA/graphene 3D-printed electrode with nickel microparticles was a great alternative for the fabrication of sensors.

Regarding the use of conductive filaments for the construction of biosensors, Marzo et al. [27] produced a 3D-printed PLA/graphene electrode using the FDM technique. The obtained device was modified with horseradish peroxidase (HPR) to develop a biosensor capable of detecting hydrogen peroxide. Thus, initially, the complete activation of the surface (chemical and electrochemical—DMF-EC) of the 3D electrodes was performed, and later, the sensors were modified with gold nanoparticles (AuNPs). An illustration of the whole process is shown in Fig. 8. In the same way, Silva et al., 2020, presented an unprecedented procedure in which they

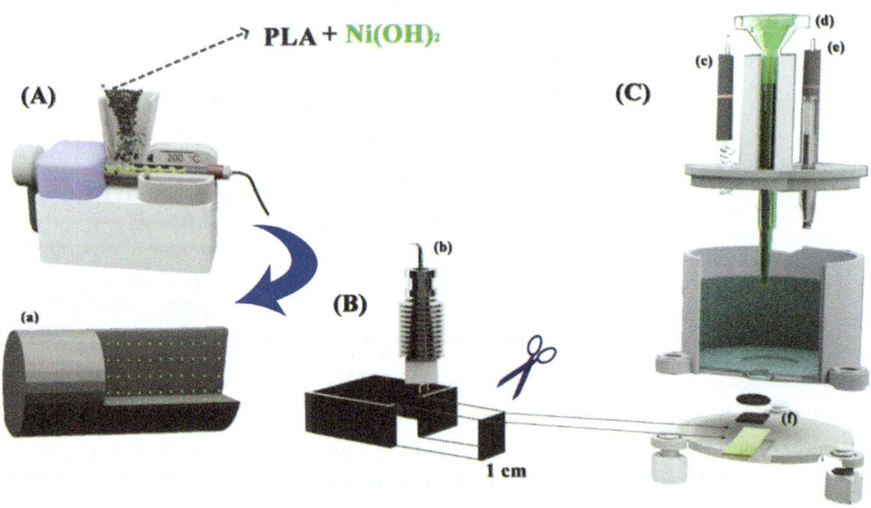

Fig. 7 Schematic diagrams: **A** Production of the Ni-G-PLA filament using the 3D extruder; **B** 3D printing of a hollow square box (4 cm × 4 cm × 2 cm) with the wall thickness of 0.72 mm; **C** the 3D-printing electrode (1 × 1 cm) is positioned at the bottom of the BIA cell on a metal plate (electrical contact); **a** illustration of Ni-G-PLA filament; **b** printer nozzle; **c** Pt counter electrode; **d** micropipette tip; **e** reference electrode (Ag/AgCl); **f** 3D-printed Ni-G-PLA working electrode (1 × 1 cm). (Reprinted from [26] with permission of Elsevier)

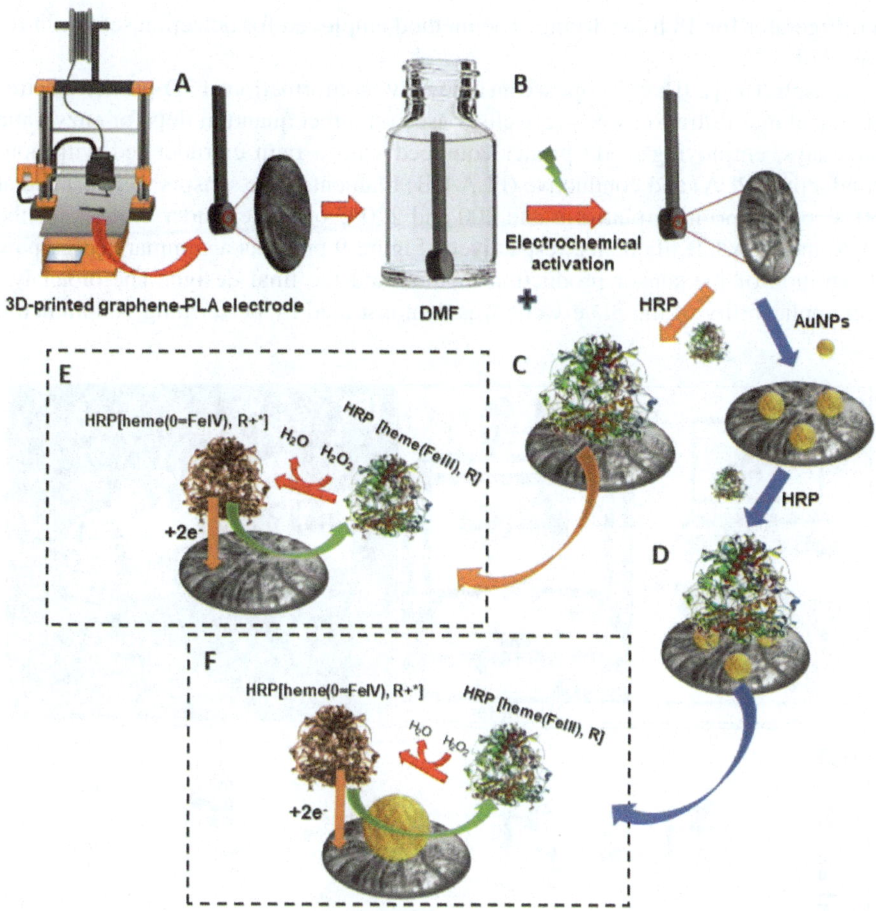

Fig. 8 Representative scheme of 3D graphene-PLA biosensor fabrication: **a** 3D printing of the electrode; **b** activation in DMF and by electrochemistry; **c** modification of 3D-printed electrode with HRP enzyme; and **d** modification of the 3D-printed electrode with gold NPs and, subsequently, with HRP enzyme. **e** and **f** are corresponding mechanisms of H_2O_2 detection. (Reprinted from [27] with permission of Elsevier)

have employed different chemical treatments to form reduced graphene oxide (rGO) in 3D-printed electrodes made from conductive filaments of PLA/graphene printed in 3D by FDM for the determination of serotonin and catechol [28]. Tyrosinase enzyme was immobilized on the surface of the working electrode to obtain a biosensor for catechol determination. Thus, to obtain the enzyme layer film, 1.0 mg of dihexadecyl phosphate (DHP) was dissolved in 1.0 mL of 0.2 mol L^{-1} phosphate buffer (pH 6.0), and 90 μL of this solution was mixed with the enzyme tyrosinase (10 μL/25 units) under constant stirring per 10 s. Next, a volume of 40 μL of this solution was placed, using a micropipette, upon the treated G-PLA surface, and the system was kept inside a refrigerator for 48 h for drying. The method employed for detection serotonin for catechol.

Katseli et al. [29] have proposed an innovative conformation of 3D-printed electrochemical microtitration wells (e-wells) based on direct quantum dots for enzymatic bioassays, employing a 3D printer equipped with a twin extruder fed with non-conductive (PLA) and conductive (PLA/CB) filaments. The sensors were printed at 60 °C on the printing platform and 200 and 220°C on the extruder nozzles for the PLA and PLA/CB filaments, respectively. Figure 9 presents a summarized graphic illustration of the sensor production method and the final design. The bioanalytical applicability of the 3D e-wells was demonstrated by performing voltammetric

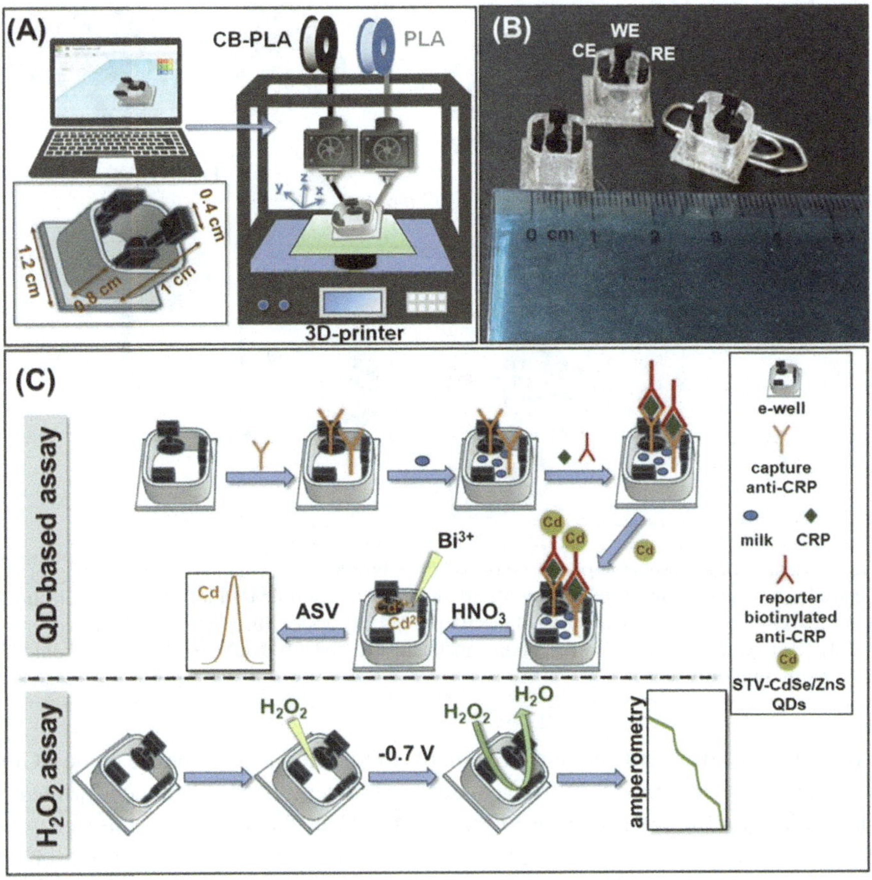

Fig. 9 **a** 3D-printing fabrication procedure of e-well and its dimension in cm. The conductive filament is PLA loaded with carbon black (PLA/CB). **b** Photograph of the 3D-printed e-wells. **c** Schematic illustration of the immunoassay for the QD-based voltammetric determination of CRP and H_2O_2 amperometric assay in 3D e-wells. (Reprinted from [29] with permission of Wiley Analytical Science)

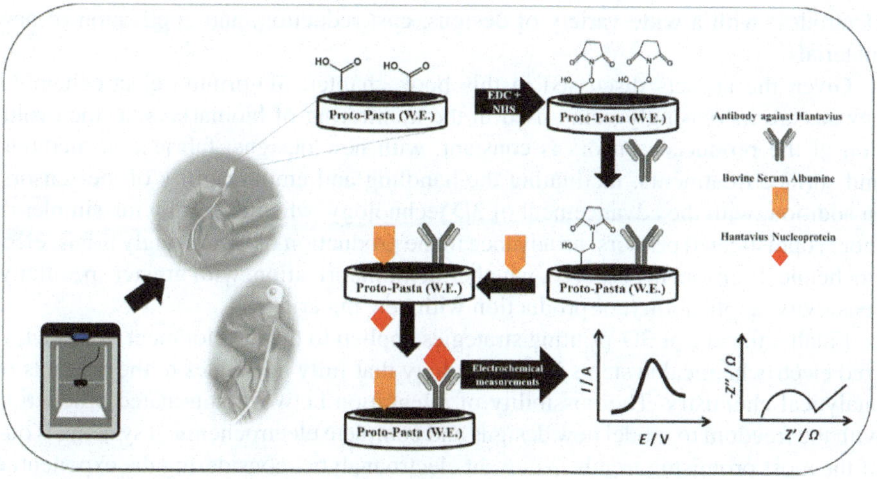

Fig. 10 Immunosensor step-by-step buildup. (Reprinted from [30] with permission of Elsevier)

bioassays in the detection of the C-reactive protein biomarker employing biotinylated reporter antibody and streptavidin-conjugated CdSe/ZnS QDs. In addition, due to the extension of its scope to the enzymatic biosensing, the e-wells were applied for the determination of by-products of hydrogen peroxide, demonstrating universal applicability in electrochemical bioassays.

Finally, Martins et al. [30] have recently reported the first immunosensor constructed with the commercial 3D conductive filament of CB/(PLA) to detect Hantavirus Araucaria nucleoprotein (Np). In this work, the biorecognition element (antibody against Hantavirus) was directly immobilized on the 3D-printed electrodes using EDC/NHS chemistry (please, see Fig. 10). To detect the hantavirus nucleoprotein (Np), the electrode's response toward the redox probe ($K_3[Fe(CN)_6]$) was compared in the absence and presence of Np. By applying this simple biosensing approach, it was possible to quantify the Hantavirus Araucaria, which was successfully applied in the analysis of diluted human serum samples.

4 Conclusions and Perspectives

The development of new types of devices for the sensing of biomarkers is of paramount importance, especially devices that aim to overcome the current problems of conventional methods of analysis. In this way, 3D-printing technology is a fundamental tool to support the production of a new generation of electrochemical devices. This technology, coupled with the development of new electrochemical sensors, brings several advantages, such as automatization and large-scale production, development of complete and miniaturized systems (cell and electrodes) with a wide variety of designs, cost reduction, and application of new materials.

Given the aspects discussed in this book chapter, 3D-printed electrochemical devices are increasingly being used in the monitoring of biomarkers, as the evolution of the produced sensors is constant, with new designs, fabrication methods, and surface treatments, facilitating the handling and employability of the sensors. In addition, with the advancement of 3D technology, whether employing simpler or more sophisticated printers, an advance in the production of increasingly robust electrochemical sensors is observed, whether in miniaturization, with greater specificity, sensitivity, applicability, or production with new materials.

Finally, the use of 3D-printing strategies applied to the development of miniaturized electrochemical systems is a technology that fully adequates to the precepts of analytical chemistry. The possibility of integration between automated production with the freedom to model new designs and complete electrochemical systems is one of the most promising trends in current electroanalytic, considering the exponential growth of publications. Thus, this manufacturing strategy will become increasingly popular and with wide growth, aiming the production of new types of materials, new methods of surface treatment, or new architectures that brings beneficial advances for both the academic environment and society.

Acknowledgements The authors are grateful to CNPq (402943/2016-3, 408309/2018-0, 303338/2019-9, 309803/2020-9, 311290/2020-5), CAPES (001 and Pandemias 8887.504861/2020-00), FAPESP (2017/21097-3), and FAPEMIG (APQ-00083-21) for the financial support.

References

1. Cardoso, R.M., Kalinke, C., Rocha, R.G., dos Santos, P.L., Rocha, D.P., Oliveira, P.R., Janegitz, B.C., Bonacin, J.A., Richter, E.M., Munoz, R.A.A.: Additive-manufactured (3D-printed) electrochemical sensors: A critical review. Anal. Chim. Acta. **1118**, 73–91 (2020). https://doi.org/10.1016/j.aca.2020.03.028
2. Omar, M.H., Razak, K.A., Ab Wahab, M.N., Hamzah, H.H.: Recent progress of conductive 3D-printed electrodes based upon polymers/carbon nanomaterials using a fused deposition modelling (FDM) method as emerging electrochemical sensing devices, RSC Adv. **11**, 16557–16571 (2021). https://doi.org/10.1039/d1ra01987b
3. Liyarita, B.R., Ambrosi, A., Pumera, M.: 3D-printed electrodes for sensing of biologically active molecules. Electroanalysis **30**, 1319–1326 (2018). https://doi.org/10.1002/elan.201700828
4. Diegel, O.: Additive manufacturing: an overview. Elsevier (2014). https://doi.org/10.1016/B978-0-08-096532-1.01000-1
5. Jiménez, M., Romero, L., Domínguez, I.A., Espinosa, M.D.M., Domínguez, M.: Additive manufacturing technologies: an overview about 3D printing methods and future prospects, complexity (2019). https://doi.org/10.1155/2019/9656938
6. Su, C.K.: Review of 3D-printed functionalized devices for chemical and biochemical analysis. Anal. Chim. Acta. **1158**, 338348 (2021). https://doi.org/10.1016/j.aca.2021.338348
7. Abdalla, A., Patel, B.A.: 3D-printed electrochemical sensors: A new horizon for measurement of biomolecules. Curr. Opin. Electrochem. **20**, 78–81 (2020). https://doi.org/10.1016/j.coelec.2020.04.009

8. Ambrosi, A., Pumera, M.: 3D-printing technologies for electrochemical applications. Chem. Soc. Rev. **45**, 2740–2755 (2016). https://doi.org/10.1039/C5CS00714C
9. Cardoso, R.M., Mendonça, D.M.H., Silva, W.P., Silva, M.N.T., Nossol, E., da Silva, R.A.B., Richter, E.M., Muñoz, R.A.A.: 3D printing for electroanalysis: from multiuse electrochemical cells to sensors, Anal. Chim. Acta. **1033**, 49–57 (2018). https://doi.org/10.1016/j.aca.2018.06.021
10. dos Santos, P.L., Katic, V., Loureiro, H.C., dos Santos, M.F., dos Santos, D.P., Formiga, A.L.B., Bonacin, J.A.: Enhanced performance of 3D printed graphene electrodes after electrochemical pre-treatment: role of exposed graphene sheets. Sens.S Actuators B Chem. **281**, 837–848 (2019). https://doi.org/10.1016/j.snb.2018.11.013
11. Foster, C.W., Down, M.P., Zhang, Y., Ji, X., Rowley-Neale, S.J., Smith, G.C., Kelly, P.J., Banks, C.E.: 3D Printed graphene based energy storage devices. Sci. Rep. **7**, 1–11 (2017). https://doi.org/10.1038/srep42233
12. dos Santos, P.L., Rowley-Neale, S.J., Ferrari, A.G.-M., Bonacin, J.A., Banks, C.E.: Ni–Fe (Oxy)hydroxide modified graphene additive manufactured (3D-printed) electrochemical platforms as an efficient electrocatalyst for the oxygen evolution reaction. ChemElectroChem. **6**, 5633–5641 (2019). https://doi.org/10.1002/celc.201901541
13. Rohaizad, N., Mayorga-Martinez, C.C., Novotný, F., Webster, R.D., Pumera, M.: 3D-printed Ag/AgCl pseudo-reference electrodes, Electrochem. Commun. **103**, 104–108 (2019). https://doi.org/10.1016/j.elecom.2019.05.010
14. Richter, E.M., Rocha, D.P., Cardoso, R.M., Keefe, E.M., Foster, C.W., Munoz, R.A.A., Banks, C.E.: Complete additively manufactured (3D-Printed) electrochemical sensing platform. Anal. Chem. **91**, 12844–12851 (2019). https://doi.org/10.1021/acs.analchem.9b02573
15. Kalinke, C., Neumsteir, N.V., Aparecido, G.D.O., Ferraz, T.V.D.B., Dos Santos, P.L., Janegitz, B.C., Bonacin, J.A.: Comparison of activation processes for 3D printed PLA-graphene electrodes: electrochemical properties and application for sensing of dopamine. Analyst. **145**, 1207–1218 (2020). https://doi.org/10.1039/c9an01926j
16. Cardoso, R.M., Silva, P.R.L., Lima, A.P., Rocha, D.P., Oliveira, T.C., do Prado, T.M., Fava, E.L., Fatibello-Filho, O., Richter, E.M., Muñoz, R.A.A.: 3D-printed graphene/polylactic acid electrode for bioanalysis: biosensing of glucose and simultaneous determination of uric acid and nitrite in biological fluids. Sens.S Actuators, B Chem. **307**, 127621 (2020). https://doi.org/10.1016/j.snb.2019.127621
17. Walters, J.G., Ahmed, S., Terrero Rodríguez, I.M., O'Neil, G.D.: Trace analysis of heavy metals (Cd, Pb, Hg) using native and modified 3D printed graphene/poly(lactic acid) composite electrodes. Electroanalysis. **32**, 859–866 (2020). https://doi.org/10.1002/elan.201900658
18. Foster, C.W., Elbardisy, H.M., Down, M.P., Keefe, E.M., Smith, G.C., Banks, C.E.: Additively manufactured graphitic electrochemical sensing platforms. Chem. Eng. J. **381**, 122343 (2020). https://doi.org/10.1016/j.cej.2019.122343
19. Vaněčková, E., Bouša, M., Sokolová, R., Moreno-García, P., Broekmann, P., Shestivska, V., Rathouský, J., Gál, M., Sebechlebská, T., Kolivoška, V.: Copper electroplating of 3D printed composite electrodes, J. Electroanal. Chem. **858**, 113763 (2020). https://doi.org/10.1016/j.jelechem.2019.113763
20. Katic, V., dos Santos, P.L., dos Santos, M.F., Pires, B.M., Loureiro, H.C., Lima, A.P., Queiroz, J.C.M., Landers, R., Muñoz, R.A.A., Bonacin, J.A.: 3D printed graphene electrodes modified with prussian blue: emerging electrochemical sensing platform for peroxide detection. ACS Appl. Mater. Interfaces. **11**, 35068–35078 (2019). https://doi.org/10.1021/acsami.9b09305
21. Liyarita, B., Ambrosi, A., Pumera, M.: 3D-printed electrodes for sensing of biologically active molecules. Electroanalysis, U. **30**, 1319–1326 (2018). https://doi.org/10.1002/elan.201700828
22. Ho, E.H.Z., Ambrosi, A., Pumera, M.: Additive manufacturing of electrochemical interfaces: simultaneous detection of biomarkers. Appl. Mater. Today. **12**, 43–50 (2018). https://doi.org/10.1016/j.apmt.2018.03.008
23. dos Santos, P.L., Katic, V., Loureiro, H.C., dos Santos, M.F., dos Santos, D.P., Formiga, A.L.B., Bonacin, J.A.: Enhanced performance of 3D printed graphene electrodes after electrochemical pre-treatment: role of exposed graphene sheets. Sens.S Actuators, B Chem. **281**, 837–848 (2019). https://doi.org/10.1016/j.snb.2018.11.013

24. Katseli, V., Economou, A., Kokkinos, C.: Single-step fabrication of an integrated 3D-printed device for electrochemical sensing applications. Electrochem. Commun. **103**, 100–103 (2019). https://doi.org/10.1016/j.elecom.2019.05.008
25. Kalinke, C., Neumsteir, N.V., Roberto de Oliveira, P., Janegitz, B.C., Bonacin, J.A.: Sensing of L-methionine in biological samples through fully 3D-printed electrodes. Anal. Chim. Acta. **1142**, 135–142 (2021). https://doi.org/10.1016/j.aca.2020.10.034
26. Rocha, R.G., Cardoso, R.M., Zambiazi, P.J., Castro, S.V.F., Ferraz, V.B., Aparecido, G.D.O., Bonacin, J.A., Munoz, R.A.A., Richter, E.M.: Production of 3D-printed disposable electrochemical sensors for glucose detection using a conductive filament modified with nickel microparticles. Anal. Chim. Acta. (2020). https://doi.org/10.1016/j.aca.2020.07.028
27. López Marzo, A.M., Mayorga-Martinez, C.C., Pumera, M.: 3D-printed graphene direct electron transfer enzyme biosensors. Biosens. Bioelectron. **151**, 111980 (2020). https://doi.org/10.1016/j.bios.2019.111980
28. Silva, V.A.O.P., Fernandes-Junior, W.S., Rocha, D.P., Stefano, J.S., Munoz, R.A.A., Bonacin, J.A., Janegitz, B.C.: 3D-printed reduced graphene oxide/polylactic acid electrodes: a new prototyped platform for sensing and biosensing applications, Biosens. Bioelectron. **170** (2020). https://doi.org/10.1016/j.bios.2020.112684
29. Katseli, V., Angelopoulou, M., Kokkinos, C.: 3D printed mioelectronic microwells. Wiley Online Libr. **31** (2021). https://doi.org/10.1002/adfm.202102459
30. Martins, G., Gogola, J.L., Budni, L.H., Janegitz, B.C., Marcolino-Junior, L.H., Bergamini, M.F.: 3D-printed electrode as a new platform for electrochemical immunosensors for virus detection, Anal. Chim. Acta. **1147**, 30–37 (2021). https://doi.org/10.1016/j.aca.2020.12.014
31. Ng, S., Iffelsberger, C., Michalička, J., Pumera, M.: Atomic layer deposition of electrocatalytic insulator Al2O3 on three-dimensional printed nanocarbons. ACS Nano **15**, 686–697 (2021). https://doi.org/10.1021/acsnano.0c06961

The manufacturer's authorised representative in the EU is Springer Nature Customer Service Centre GmbH, Europaplatz 3, 69115 Heidelberg, Germany. If you have any concerns regarding our products, please contact ProductSafety@springernature.com

Printed and bound by CPI Group (UK) Ltd, Croydon, CR0 4YY
30/09/2026
02225309-0005